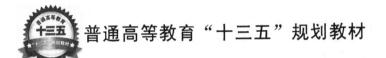

Android 应用程序开发教程
——Android Studio 版

罗文龙　主编

邹　洋　　王咪琳　　黄天祥　副主编

电子工业出版社

Publishing House of Electronics Industry

北京·BEIJING

内 容 简 介

随着移动互联网的快速发展，作为占据移动互联网半壁江山的 Android 也迎来了更大的发展机遇，与其他行业类比，移动互联网时代的 Android 发展潜力无疑是最大的(据 2016 年最新数据统计，目前，Android 的市场占有率为 63.8%，iOS 占 19.1%)。

本书基于 Google 最新推出的 Android IDE - Android Studio 和最新的 Android 模拟器 Genymotion 作为开发环境进行编写，全面介绍了 Android 应用开发的相关知识，内容覆盖了 Android 系统与开发环境、Android UI 设计、基本程序单元 Activity、Android 应用核心 Intent 和 Filters、Android 事件处理、Android 服务、Android 广播接收器、ContentProvider 实现数据共享、图形图像与多媒体、Android 的网络编程基础、Android 数据存储、GPS 应用开发与百度地图等。

本书不局限于介绍 Android 编程的各种理论知识，而是以"实例驱动"的方式来讲解。全书共 50 多个实例，这些实例能帮助读者更好地理解 Android 各知识在实际开发中的应用方式。本书最后还提供了 4 个综合项目案例：基于 Android 的计算器、图形化数字游戏、基于 Android 的音频播放器和 Android 数据库记事本，综合运用前面的各项知识点，具有较高的参考价值。与本书配套的所有实例和综合项目都可以登录华信教育资源网 www.hxedu.com.cn 免费下载。

本书可作为高等院校计算机科学与技术、软件工程、信息管理、电子商务等相关专业本科生和研究生教材，也可供从事移动开发的工作者学习参考。

未经许可，不得以任何方式复制或抄袭本书之部分或全部内容。
版权所有，侵权必究。

图书在版编目（CIP）数据

Android 应用程序开发教程：Android Studio 版／罗文龙主编．—北京：电子工业出版社，2016.8
普通高等教育"十三五"规划教材
ISBN 978-7-121-28930-9

I．①A… II．①罗… III．①移动终端－应用程序－程序设计－高等学校－教材 IV．①TN929.53

中国版本图书馆 CIP 数据核字（2016）第 117299 号

策划编辑：秦淑灵
责任编辑：郝黎明　　特约编辑：张燕虹
印　　刷：北京虎彩文化传播有限公司
装　　订：北京虎彩文化传播有限公司
出版发行：电子工业出版社
　　　　　北京市海淀区万寿路 173 信箱　邮编：100036
开　　本：787×1 092　1/16　印张：20.75　字数：531 千字
版　　次：2016 年 8 月第 1 版
印　　次：2021 年 8 月第 9 次印刷
定　　价：42.00 元

凡所购买电子工业出版社图书有缺损问题，请向购买书店调换。若书店售缺，请与本社发行部联系，联系及邮购电话：(010)88254888，88258888。
质量投诉请发邮件至 zlts@phei.com.cn，盗版侵权举报请发邮件至 dbqq@phei.com.cn。
本书咨询联系方式：(010)88254531。

前　　言

　　移动互联网如潮水一般席卷着全世界，无论是个人还是企业、工作或生活，都极大地受其影响。移动互联网时代已经开启，它已成为全世界商业和科技创新发展的加速器，成为当下时代最大的机遇和挑战。

　　Android 系统就是一个开放式的移动互联网操作系统。今天，Android 已经成为移动互联网的宠儿，是应用最广泛的移动互联网平台（据 2016 年最新数据统计，目前，Android 的市场占有率为 63.8%，iOS 占 19.1%）。因此，手机软件在当今的 IT 行业中具有举足轻重的地位。从招聘市场的情况来看，Android 软件人才的需求也越来越大。

　　在 2013 年 Google I/O 大会上，Google 正式推出了官方 Android 软件集成开发工具 Android Studio，并在 2015 年宣布停止对 Android Eclipse Tools 的支持。以前，很多书籍都是以 Eclipse 为开发环境进行编写的，但以后 Android Studio IDE 开发必是大势所趋，所以本书采用 Android Studio 作为实例开发平台进行讲解。

　　本书注重讲解手机应用开发的最新技术和最新应用，突出先进性、系统性、实用性和可操作性，能够使读者在较短的时间内进行 Android 开发环境的搭建，深刻理解 Android 平台体系结构，熟练使用 Android 基本组件、Android 的存储操作、多媒体开发、网络应用程序开发等技术，从而具备基本的算法设计能力、一定的系统设计和模块设计能力、一定的需求分析能力和软件代码编写能力。本教材具有以下特色。

1. 全新的开发环境

　　本书以最新的 Google 官方 Android IDE-Android Studio V1.3 为开发环境对实例进行开发讲解，让读者更快地了解 Android Studio 的界面操作。同时，还引入了当前应用广泛的 Android 模拟器软件 Genymotion，并对 Genymotion 的安装、使用进行了详细介绍，让开发者摆脱 Android 模拟器运行缓慢、耗内存的缺点，使学习 Android 开发更加得心应手。

2. 由浅入深，循序渐进

　　本书以高等教育本科生为对象，从了解 Android 和搭建开发环境学起，再学习 Android 开发的基础技术，然后学习 Android 开发的高级内容，最后学习如何开发一个完整项目。讲解过程中步骤详尽、版式新颖，并在操作的内容图片上进行了标注，让读者在阅读时一目了然，从而快速掌握书中内容。

3. 知识全面、覆盖面广

　　书中全方位介绍了 Android 的相关知识：Android 系统与开发环境、Android UI 设计、基本程序单元 Activity、Android 应用核心 Intent 和 Filters、Android 事件处理、Android 服务、Android 广播接收器、ContentProvider 实现数据共享、图形图像与多媒体、Android 网络编程基础、数据存储、GPS 应用开发与百度地图等。

4．实例典型，轻松易学

通过实例进行学习是最好的学习方式，本书通过"实例驱动"的方式来讲解理论知识。全书共 50 多个实例，这些实例能帮助读者更好地理解 Android 各知识在实际开发中的应用方式。本书最后还提供了 4 个综合项目案例：基于 Android 的计算器、图形化数字游戏、基于 Android 的音频播放器和 Android 数据库记事本，综合运用前面的各项知识点，具有较高的参考价值。

本书可作为高等院校计算机科学与技术、软件工程、信息管理、电子商务等相关专业本科生和研究生教材，也可供从事移动开发工作者学习参考。与本书配套的所有实例和综合项目都可以登录华信教育资源网 www.hxedu.com.cn 免费下载。

本书由罗文龙担任主编，邹洋、王咪琳、黄天祥担任副主编。罗文龙执笔编写第 1~8 章，邹洋编写第 13~16 章，王咪琳、黄天祥编写第 9~12 章，全书由罗文龙负责审校和统稿。

特别感谢蹇洁教授对本书编写提供的指导和大力支持。同时感谢本书的编辑，没有她们的策划、指导、无私帮助和辛勤工作，就不会有本书的出版。

本书仅以 Android Studio V1.3 + API 18/23 + Genymotion 为开发环境进行讲解，书中所论并不完美，错误和疏漏之处，恳请读者批评指正。笔者 E-mail：luowl@cqupt.edu.cn。

编　者
2016 年 4 月

目 录

第1篇 基 础 篇

第1章 Android 系统与开发环境 2
1.1 什么是 Android 2
1.1.1 Android 平台的架构 2
1.1.2 Android 操作系统的功能 3
1.1.3 Android 的分支平台 5
1.1.4 Android 平台五大优势特色 5
1.2 Android Studio 入门 6
1.2.1 Android Studio 的特点 6
1.2.2 Android Studio 系统要求 6
1.2.3 Android Studio 和 Eclipse ADT 比较 7
1.3 搭建 Android 开发环境 7
1.3.1 下载 JDK 7
1.3.2 JDK 的安装 8
1.3.3 JDK 的环境变量配置 10
1.3.4 Windows 环境下安装 Android Studio 13
1.3.5 Mac 环境下安装 Android Studio 15
1.3.6 下载、安装和配置 Android SDK 16
1.3.7 SDK 下载国内镜像 20
1.3.8 第一次打开 Android Studio 的配置 21
1.4 Android 模拟器 Genymotion 23
1.4.1 什么是 Genymotion 23
1.4.2 Genymotion 的特性 23
1.4.3 Genymotion 运行环境要求 24
1.4.4 注册和下载 Genymotion 模拟器 24
1.4.5 安装 Genymotion 模拟器 27
1.4.6 使用 Genymotion 模拟器 28
1.4.7 给 Android Studio 安装 Genymotion 插件 31
1.5 新建一个 Hello World 项目 33

第2章 Android UI 设计 38
2.1 Android UI 布局 38
2.2 ListView（列表视图） 46

		2.2.1	ListView（列表视图）的常用属性	46
		2.2.2	ArrayAdapter（数组适配器）	47

2.3 GridView（网格视图） 49
2.4 Android UI 控件 56
    2.4.1 TextView（文本框） 56
    2.4.2 EditText（编辑框） 59
    2.4.3 AutoCompleteTextView（自动填充文本框） 61
    2.4.4 Button（普通按钮） 64
    2.4.5 ImageButton（图片按钮） 66
    2.4.6 CheckBox（复选框） 69
    2.4.7 ToggleButton（开关按钮） 73
    2.4.8 RadioButton（单选按钮）与 RadioGroup（按钮组） 77
    2.4.9 使用 ProgressDialog（进度对话框）类创建 ProgressBar（进度条） 81
    2.4.10 Spinner（列表选择框） 84
    2.4.11 TimePicker（时间拾取器） 87
2.5 DatePicker（日期拾取器）与 DatePickerDialog（日期拾取器对话框） 91

第 3 章 基本程序单元 Activity 97

第 4 章 Android 应用核心 Intent 和 Filters 102
4.1 Intent 的各属性 102
    4.1.1 Component（组件） 103
    4.1.2 Action（动作） 103
    4.1.3 Category（类别） 104
    4.1.4 Data（数据）和 Type（类型） 105
    4.1.5 Extra（额外） 106
    4.1.6 Flag（标记） 106
4.2 Intents 的类型 107
    4.2.1 显式 Intent 107
    4.2.2 隐式 Intent 107
4.3 Intent Filters（意图过滤器） 111

第 5 章 Android 事件处理 119
5.1 Android 事件处理概述 119
5.2 注册事件监听器的方法 119

第 6 章 Android 服务 129
6.1 Service 的分类 129
6.2 Service 的生命周期 129
6.3 Service 生命周期中的回调方法 130

第 7 章 Android 广播接收器 135
7.1 创建 Broadcast Receiver 135

7.2	注册 Broadcast Receiver	135
	7.2.1 接收用户自定义 Broadcast Intent 消息	136
	7.2.2 接收系统广播消息	136

第 8 章 ContentProvider 实现数据共享 — 144

- 8.1 ContentProvider 概述 — 144
- 8.2 URI 简介 — 145
- 8.3 创建 ContentProvider — 145

第 9 章 图形图像与多媒体 — 156

- 9.1 基础绘图 — 156
  - 9.1.1 常用绘图类 — 156
  - 9.1.2 绘制 2D 图形 — 157
- 9.2 位图操作 — 160
- 9.3 Android 中的动画 — 163
  - 9.3.1 Frame Animation — 163
  - 9.3.2 Tween Animation — 165
  - 9.3.3 Property Animation — 173
  - 9.3.4 动画监听器：AnimationListener — 173
- 9.4 在 Android 中播放音频与视频 — 173
  - 9.4.1 MediaPlayer 介绍 — 173
  - 9.4.2 运用 MediaPlayer 播放音频 — 176
  - 9.4.3 播放视频 — 181
- 9.5 控制相机拍照 — 187

第 10 章 Android 的网络编程基础 — 189

- 10.1 基于 TCP 协议的网络通信 — 189
  - 10.1.1 TCP/IP 协议基础 — 189
  - 10.1.2 Socket 与 ServerSocket 建立通信 — 190
- 10.2 使用 URL 访问网络 — 193
  - 10.2.1 使用 URL 获取网络资源 — 193
  - 10.2.2 使用 URLConnection 提交请求 — 196
  - 10.2.3 使用 HttpURLConnection 实现网络通信 — 201
- 10.3 使用 WebView — 203
  - 10.3.1 使用 WebView 浏览网页 — 203
  - 10.3.2 加载本地 html 页面 — 205
  - 10.3.3 JavaScript 交互调用 — 207
- 10.4 使用 WebService 进行网络编程 — 209
  - 10.4.1 WebService 基础 — 209
  - 10.4.2 调用 WebService — 210
  - 10.4.3 实现手机归属地查询 — 212

第 11 章　Android 数据存储 · 216

11.1　使用 SharedPreferences · 216
11.1.1　SharedPreferences 的使用方法 · 216
11.1.2　SharedPreferences 的应用 · 217
11.2　File 存储 · 219
11.2.1　使用 IO 流操作文件 · 219
11.2.2　文件操作应用 · 219
11.2.3　将文件保存到 SD 卡 · 221
11.3　SQLite 数据库 · 224
11.3.1　SQLite 介绍 · 224
11.3.2　SQLite 数据库操作 · 227

第 12 章　GPS 应用开发与百度地图 · 238
12.1　支持 GPS 的核心 API · 238
12.2　获取 LocationProvider · 239
12.3　获取定位信息 · 241
12.4　使用百度地图服务 · 244
12.4.1　获取百度地图 SDK · 244
12.4.2　使用百度地图定位 · 246

第 2 篇　综合项目篇

第 13 章　基于 Android 的计算器 · 252
13.1　系统分析 · 252
13.2　系统设计 · 252
13.2.1　系统目标 · 252
13.2.2　系统功能结构 · 252
13.2.3　系统业务流程 · 253
13.3　系统实施 · 253
13.3.1　开发及运行环境 · 253
13.3.2　项目的创建 · 254
13.3.3　计算器主界面的实现 · 256
13.3.4　计算器逻辑的实现 · 259
13.4　系统运行与测试 · 264

第 14 章　图形化数字游戏 · 265
14.1　系统分析 · 265
14.2　系统设计 · 265
14.2.1　系统目标 · 265
14.2.2　系统功能结构 · 265
14.2.3　系统业务流程 · 266
14.3　系统实施 · 266

  14.3.1 开发及运行环境 266
  14.3.2 项目的创建 266
  14.3.3 项目工程结构 269
  14.3.4 游戏主界面的实现 269
  14.3.5 数字小游戏逻辑的实现 271
 14.4 系统运行与测试 281

第 15 章 基于 Android 的音频播放器 283
 15.1 系统分析 283
 15.2 系统设计 283
  15.2.1 系统目标 283
  15.2.2 系统功能结构 283
  15.2.3 系统业务流程 284
 15.3 系统实施 284
  15.3.1 开发及运行环境 284
  15.3.2 项目的创建 285
  15.3.3 项目工程结构 288
  15.3.4 音频播放器主界面的实现 288
  15.3.5 Android 音频播放器逻辑的实现 292
 15.4 系统运行与测试 298

第 16 章 Android 数据库记事本 299
 16.1 系统分析 299
 16.2 系统设计 299
  16.2.1 系统目标 299
  16.2.2 系统功能结构 299
  16.2.3 系统业务流程 300
 16.3 系统实施 300
  16.3.1 开发及运行环境 300
  16.3.2 项目的创建 300
  16.3.3 项目工程结构 304
  16.3.4 记事本主界面的实现 304
  16.3.5 记事本逻辑的实现 308
 16.4 系统运行与测试 318

参考文献 320

第 1 篇

基 础 篇

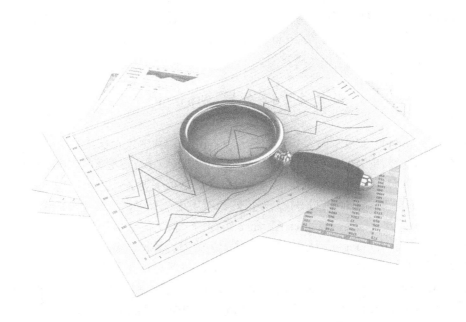

第 1 章　Android 系统与开发环境
第 2 章　Android UI 设计
第 3 章　基本程序单元 Activity
第 4 章　Android 应用核心 Intent 和 Filters
第 5 章　Android 事件处理
第 6 章　Android 服务
第 7 章　Android 广播接收器
第 8 章　ContentProvider 实现数据共享
第 9 章　图形图像与多媒体
第 10 章　Android 的网络编程基础
第 11 章　Android 数据存储
第 12 章　GPS 应用开发与百度地图

第 1 章　Android 系统与开发环境

　　黑莓和 iPhone 都提供了受欢迎、高容量的移动平台，但却分别针对两个不同的消费群体。黑莓是企业业务用户的不二选择。但是，作为一种消费设备，它在易用性和新奇性方面难以与 iPhone 抗衡。Android 系统则是一个成熟的、全开放的平台，它具有潜力且涵盖移动电话的两个不同消费群体，甚至可能缩小工作和娱乐之间的差别。Android 系统是一个可靠的平台，它可以经济有效地进行安装部署和提供支持，并且具有良好的设计、开发环境。如今，Android 系统已经成为全球应用最广泛的手机操作系统。三星、HTC、华为和小米手机通过 Android 平台获得了巨大的成功，企业对 Android 开发人才的需求也迅猛增长。Android 平台上的移动开发在软件开发的地位将显得越来越重要。

1.1　什么是 Android

　　Android 是 Google 公司专门为移动设备开发的平台，其中包含操作系统、中间件和核心应用等。Android 最初由 Andy Rubin（Android 之父）创办。Google 公司于 2005 年收购了成立约 22 个月的 Android 公司，开始了短信、手机检索、定位等业务，进入了基于 Linux 平台的开发。Google 公司在 2007 年 11 月 5 日正式公布了这个平台，之后由开放手机联盟（Open Handset Alliance）开发。Open Handset Alliance 组织由一群共同致力于构建更好的移动电话的公司组成。这个组织由 Google 领导，包含了移动运营商、手持设备制造商、零部件制造商、软件解决方案和平台提供商以及市场营销公司。Android 平台的更新速度很快，第一个版本是 Android 1.5 Cupcake（蛋糕杯），接下来是 Android 1.6 Donut（甜甜圈）、Android 2.0/2.0.1/2.1 Eclair（法式奶油夹心松饼）、Android 2.2/2.2.1 Froyo（冻酸奶）、Android 2.3 Gingerbread（姜饼）、Android 3.0 Honeycomb（蜂巢）等。截至 2016 年，Android 的版本已经到了 Android 6.0 Marshmallow（棉花糖）。从 Android 的发展来看，它已经不再局限于移动应用领域了，它还包含 Google TV（运行 Android 系统的电视机）、Google Auto（运行 Android 系统的汽车）、Google Wear（Google 穿戴设备）、Google Glass（Google 眼镜）等设备。

1.1.1　Android 平台的架构

　　从图 1.1 中可以清楚地看出 Android 平台由 Application、Application Framework、Libraries、Linux Kernel 四层构成。

1. Application

　　第一层是面向用户的 Application（应用程序），包括电话程序、联系人程序、浏览器、地图、E-mail 等。

2. Application Framework

　　第二层是 Application Framework，为开发人员提供访问核心组件所使用的 API，包含丰富的 View 组件、Content Providers（内容提供器，用来访问另一个应用程序的数据，如联系人数据库或者共享它们自己的数据)、Resource Manager(资源管理器,提供对资源的访问,如本地字符串、图形、布局文件等)、Notification Manager(通知管理器)、Activity Manager(活动管理器)等。

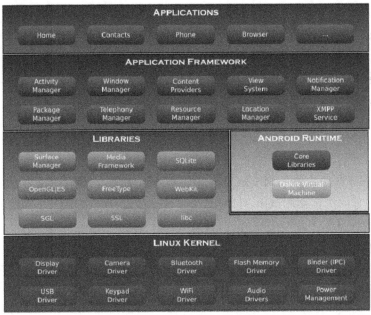

图 1.1　Android 平台的架构

3. Libraries

第三层是系统运行库,包含程序库和 Android 运行库。程序库中包含了一些 C/C++库,通过应用框架为开发者提供服务,如 Libc(从 BSD 继承来的标准 C 系统函数库)、媒体库、SGL(2D 图形引擎)、3D libraries(基于 OpenGL ES 1.0 APIs 实现,可以使用硬件或软件 3D 加速)、SQLite(数据库引擎)等。

Android 运行库还包括一个核心库,该核心库提供了 Java 编程语言核心库的大多数功能。每一个 Android 应用程序都在自己的进程中运行,都拥有一个独立的 Dalvik 虚拟机实例。Dalvik 被设计成一个设备可以同时高效地运行多个虚拟系统。Dalvik 虚拟机执行.dex 文件(Dalvik 的可执行文件),该格式文件针对小内存使用做了优化。同时,虚拟机是基于寄存器的,所有类都经由 Java 编译器编译,然后通过 SDK 中的 dx 工具转化成.dex 格式,由虚拟机执行。Dalvik 虚拟机依赖于 Linux 内核的一些功能,比如线程机制和底层内存管理机制。

4. Linux Kernel

第四层是 Linux 核心,Android 的核心系统依赖于 Linux 内核,如安全性、内存管理、进程管理、网络协议栈和驱动模型。Android 的 Linux Kernel 也不是 GNU/Linux。Android 将驱动程序移到 Userspace,使得 Linux Driver 与 Linux Kernel 分开。

1.1.2　Android 操作系统的功能

- 显示布局

Android 操作系统支持更大的分辨率,VGA、2D 显示、3D 显示都给予 OpenGL ES 3.0 标准规格(4.3 版本开始支持 OpenGL ES 3.0),并且支持传统的智能手机。

- 数据存储

Android 操作系统内置 SQLite 小型关联式数据库管理系统来负责存储数据。

- 网络

Android 操作系统支持所有的网络制式，包括 GSM/EDGE、IDEN、CDMA、TD-SCDMA、EV-DO、UMTS、Bluetooth、Wi-Fi、LTE、NFC 和 WiMAX。

- 信息

作为设计给智能手机使用的操作系统，Android 操作系统支持短信和邮件，并且支持所有的云信息和服务器信息。

- 语言

Android 操作系统支持多语言。

- 浏览器

Android 操作系统中内置的网页浏览器基于 WebKit 核心，并且采用了 Chrome V8 引擎。在 Android 4.0 内置的浏览器测试中，HTML5 和 Acid3 故障处理中均获得了满分。Android 2.2～4.0 版支持 Flash，4.0 版本之后不再支持 Flash。

- 支持 Java

虽然 Android 操作系统中的应用程序大部分都是由 Java 编写的，但是 Android 却是以转换为 Dalvik Executables 的文件在 Dalvik 虚拟机上运行的。由于 Android 中并不自带 Java 虚拟机，因此无法直接运行 Java 程序。不过，Android 平台上提供了多个 Java 虚拟机供用户下载使用，安装了 Java 虚拟机的 Android 系统可以运行 Java_ME 的程序。5.0 版(Lolipop)开始以 Android Runtime(ART)取代 Dalvik 虚拟机。

- 媒体支持

Android 操作系统本身支持以下格式的音频/视频/图片媒体：WebM、H.263、H.264(in 3GP or MP4 container)、MPEG-4 SP、AMR、AMR-WB(in 3GP container)、AAC、HE-AAC(in MP4 or 3GP container)、MP3、MIDI、Ogg Vorbis、FLAC、WAV、JPEG、PNG、GIF、BMP。如果用户需要播放更多格式的媒体，可以安装其他第三方应用程序。

- 流媒体支持

Android 操作系统支持 RTP/RTSP(3GPP PSS、ISMA)的流媒体以及(HTML5 <video>)的流媒体，同时还支持 Adobe 的 Flash，在安装了 RealPlayer 之后，还支持苹果公司的流媒体。

- 硬件支持

Android 操作系统支持识别且使用视频/照片摄像头、多点电容/电阻触摸屏、GPS，加速计、陀螺仪、气压计、磁力仪(高斯计)、键盘、鼠标、USB Disk、专用的游戏控制器、体感控制器、游戏手柄、蓝牙设备、无线设备、感应和压力传感器、温度计，加速 2D 位块传输(硬件方向、缩放、像素格式转换)和 3D 图形加速。

- 多点触控

Android 支持本地的多点触控，在最初的 HTC Hero 智能手机上即有这个功能。该功能是内核级别(为了避免对苹果公司的触屏技术造成侵权)。

- 蓝牙

Android 支持 A2DP、AVRCP、发送文件(OPP)、访问电话簿(PBAP)、语音拨号和发送智能手机之间的联系。同时，支持键盘、鼠标和操纵杆(HID)。

- 多任务处理

Android 操作系统支持本地的多任务处理。

- 语音功能

除了支持普通的电话通话之外，Android 操作系统从最初的版本开始就支持使用语音操作来使用 Google 进行网页搜索等功能。而从 Android 2.2 开始，语音功能还可以用来输入文字、语音导航等功能。

- 无线共享功能

Android 操作系统支持用户使用本机充当"无线路由器"，并且将本机的网络共享给其他智能手机，其他机器只需要通过 WiFi 查找到共享的无线热点，就可以上网。而在 Android 2.2 之前的操作系统则需要通过第三方应用或者其他定制版系统来实现这个功能。

- 截图功能

从 Android 4.0 开始，Android 操作系统便支持截图功能，该功能允许用户直接抓取智能手机屏幕上的任何画面，用户还可以通过编辑功能对截图进行处理，还可以通过蓝牙、E-mail、微博、共享等方式发送给其他用户或者上传到网络上，也可以复制到计算机中。

1.1.3 Android 的分支平台

- Android Wear：Android Wear 是专为智能手表等可穿戴式设备所设计的一个 Android 系统分支。
- Android TV：Android TV 是专为家用电视所设计的一个 Android 系统分支。
- Android Auto：Android Auto 是专为汽车所设计的一个 Android 系统。
- Android Glass：Android Glass 是专为 Google 眼镜所设计的一个 Android 系统分支。

1.1.4 Android 平台五大优势特色

- 开放性

在优势方面，Android 平台首先就是其开放性，开放的平台允许任何移动终端厂商加入到 Android 联盟中。显著的开放性可以使其拥有更多的开发者，随着用户和应用的日益丰富，一个崭新的平台也将很快走向成熟。开发性对于 Android 的发展而言，有利于积累人气，这里的人气包括消费者和厂商，而对于消费者来讲，最大的受益正是丰富的软件资源。开放的平台也会带来更大竞争，如此一来，消费者将可以用更低的价位购得心仪的手机。

- 挣脱运营商的束缚

在过去很长的一段时间里，特别是在欧美地区，手机应用往往受到运营商的制约，使用什么功能接入什么网络，几乎都受到运营商的控制。自 iPhone 上市以来，用户可以更加方便地连接网络，运营商的制约减少。随着 EDGE 和 HSDPA 这些 2G 至 3G 移动网络的逐步过渡和提升，手机随意接入网络已不是运营商口中的笑谈。

- 丰富的硬件选择

这一点还是与 Android 平台的开放性相关，由于 Android 的开放性，众多的厂商会推出千奇百怪、功能特色各具的多种产品。功能上的差异和特色不会影响到数据同步甚至软件的兼容。就好比你从诺基亚 Symbian 风格手机一下改用苹果 iPhone，同时还可将 Symbian 中优秀的软件带到 iPhone 上使用，联系人等资料也可以方便地转到 iPhone 上。

- 不受任何限制的开发商

由于 Android 平台提供给第三方开发商一个十分宽泛、自由的环境，因此不会受到各种条条框框的阻挠。可想而知，会有多少新颖别致的软件诞生。但也有其两面性，如何控制血腥、暴力、情色方面的程序和游戏是留给 Android 的难题之一。

- 无缝结合的 Google 应用

如今叱咤互联网的 Google 已经走过 10 多年历史。从搜索巨人到全面的互联网渗透，Google 服务如地图、邮件、搜索等已经成为连接用户和互联网的重要纽带，而 Android 平台手机将无缝结合这些优秀的 Google 服务。

1.2 Android Studio 入门

Android Studio 是一套由 Google 公司推出的以 IntelliJ IDEA 为基础的 Android 集成开发环境，目前已经免费向 Google 及 Android 的开发人员发放。为了简化 Android 的开发力度，Google 决定将重点建设 Android Studio 工具。Google 会在 2015 年年底停止支持其他集成开发环境，比如 Eclipse。

Android Studio 是第一个官方的 Android 开发环境。其他工具，例如 Eclipse，在 Android Studio 发布之前已经有了大规模的使用。为了帮助开发者转向 Android Studio，Google 已经写出一套迁移指南。Google 同时也发布声明称，在接下来的时间里，他们会为 Android Studio 增加一些性能工具，Eclipse 里现有的 Android 工具也会通过 Eclipse 基金会继续支持下去。

1.2.1 Android Studio 的特点

- 可视化布局：WYSIWYG 编辑器、实时编码、实时程序界面预览。
- 开发者控制台：优化提示、协助翻译、来源跟踪、宣传和营销曲线图、使用率度量。
- 基于 Gradle 的构建支持。
- Android 特定代码重构和快速修复。
- Lint 提示工具更好地对程序性能、可用性、版本兼容和其他问题进行控制捕捉。
- 支持 ProGuard 和应用签名功能。
- 基于模板的向导来生成常用的 Android 应用设计和组件。
- 自带布局编辑器，可让开发者拖放 UI 组件，并预览在不同尺寸设备上的 UI 显示效果，等等。
- 支持构建 Android Wear、TV 和 Auto 应用。
- 内置 Google Cloud Platform，支持 Google Cloud Messaging 和 App Engine 的集成。

1.2.2 Android Studio 系统要求

Android Studio 系统要求如表 1.1 所示。

表 1.1 Android Studio 系统要求

	Windows	OS X	Linux
操作系统版本	Microsoft Windows 10/8.1/8/7/ Vista/ 2003 (32 或 64 位)	OS X 10.8.5 或更高版本，最高 10.10.5 (Yosemite)	GNOME、KDE、Unity desktop on Ubuntu、Fedora、GNU/Linux Debian
内存	最低 2 GB，推荐 4 GB 内存		

	Windows	OS X	Linux
磁盘空间	500 MB 磁盘空间		
Space for Android SDK	至少 1 GB 用于 Android SDK，模拟器系统映像和缓存		
JDK 版本	Java Development Kit（JDK）7 或更高版本		
屏幕分辨率	最低 1280×800 屏幕分辨率		

1.2.3 Android Studio 和 Eclipse ADT 比较

Android Studio 和 Eclipse ADT 比较如表 1.2 所示。

表 1.2 Android Studio 和 Eclipse ADT 比较

特性	Android Studio	Eclipse ADT
编译系统	Gradle	Ant
基于 Maven 的构建依赖	是	否
构建变体和多 APK 生成	是	否
高级的 Android 代码完成和重构	是	否
图形布局编辑器	是	是
APK 签名和密钥库管理	是	是
NDK 支持	Beta	是

1.3 搭建 Android 开发环境

1.3.1 下载 JDK

Java Development Kit（JDK）是 Sun 公司（2009 年，Sun 被 Oracle 收购）针对 Java 开发人员发布的免费软件开发工具包（SDK，Software Development Kit）。自从 Java 推出以来，JDK 已经成为使用最广泛的 Java SDK。作为 Java 语言的 SDK，普通用户并不需要安装 JDK 来运行 Java 程序，只需要安装 JRE（Java Runtime Environment）。而程序开发者必须安装 JDK 来编译、调试程序。下面以目前最新的版本 JDK 8 为例，介绍下载 JDK 的方法，具体步骤如下。

（1）打开浏览器，在地址栏中输入 http://www.oracle.com/index.html，进入 Oracle 的官方主页，如图 1.2 所示。

图 1.2 Oracle 的官方主页

(2)选择"Downloads"选项卡,选择"Java for Developers",跳转页面,单击"Java Platform (JDK) 8u73 / 8u74"图标,如图 1.3 所示。

图 1.3 Java for Developers 页面

(3)在新页面中,同意协议并根据计算机硬件和操作系统选择适当的版本进行下载,如图 1.4 所示。

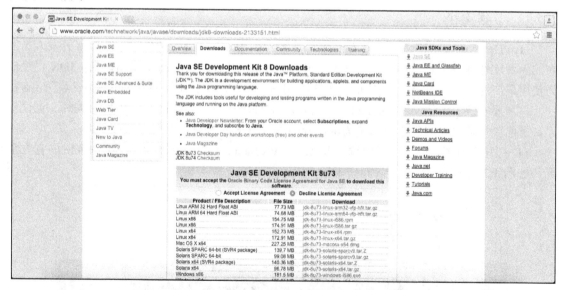

图 1.4 JDK 下载页面

1.3.2 JDK 的安装

下载完适合自己操作系统的 JDK 版本以后,就可以进行安装了。下面以 Windows 系统为例,讲解 JDK 的安装步骤。

(1)用鼠标左键双击 JDK 安装包,会出现如图 1.5 所示的对话框,然后单击"下一步"按钮。

(2)在打开的如图 1.6 所示的对话框中,单击"更改"按钮,将安装位置改为 C:\Java\jdk1.7.0_01\(根据版本决定最后的 jdk 编号),如图 1.7 所示。

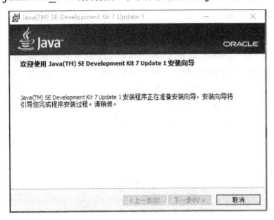

图 1.5 JDK 安装向导对话框

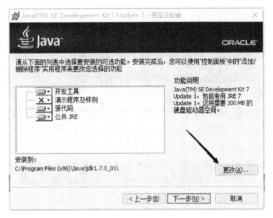

图 1.6 JDK 安装功能及位置选择对话框(一)

(3)单击"下一步"按钮,出现 JDK 正在安装界面,如图 1.8 所示。

图 1.7 JDK 安装功能及位置选择对话框(二)

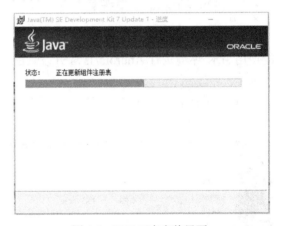

图 1.8 JDK 正在安装界面

(4)弹出如图 1.9 所示的对话框,单击"更改"按钮,将安装路径改为 C:\Java\jre7\,如图 1.9 所示。

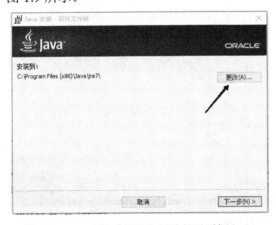

图 1.9 JRE 安装功能及位置选择对话框(一)

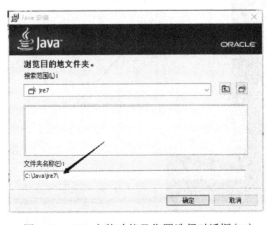

图 1.10 JRE 安装功能及位置选择对话框(二)

(5)单击"下一步"按钮进行安装,如图 1.11 所示。

(6)安装完成后,弹出如图 1.12 所示的界面,单击"完成"按钮,结束安装。

图 1.11 JRE 正在安装界面

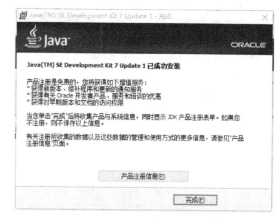

图 1.12 JDK 完成安装界面

1.3.3 JDK 的环境变量配置

完成前面的步骤,只是完成了 JDK 环境安装。这时还要通过一系列的环境变量的配置才能使用 JDK 环境进行 Android/Java 开发。配置环境变量包括 Java_home、path 和 classpath 三个部分。

(1)用鼠标右击"我的电脑",选择"属性"→"高级"→"环境变量"→"系统变量"→"新建",如图 1.13 所示。

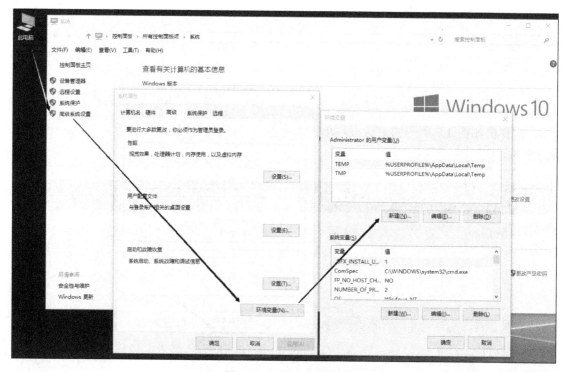

图 1.13 环境变量配置

第 1 章 Android 系统与开发环境

（2）在"变量名"输入框中输入"Java_home"，在"变量值"输入框中输入"C:\Java\jdk1.7.0_01"（根据安装路径填写），然后单击"确定"按钮，Java_home 就配置完成了。

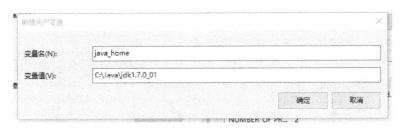

图 1.14　配置 Java_home 变量

（3）开始 classpath 变量的配置。选中"系统变量"查看是否有 classpath 项目。如果没有，就单击"新建"按钮；如果已经存在，就选中 classpath 选项，单击"编辑"按钮，然后在"变量名"框中填写"classpath"，在"变量值"框中添加"C:\Java\jdk1.7.0_01\jre\lib"（根据安装路径填写）。注意：添加部分与前面部分用"；"号隔开，如图 1.15 所示。

图 1.15　配置 classpath 变量

（4）进行 path 变量的配置。与配置"classpath"类似，在"变量名"输入框中填写"path"，在"变量值"输入框中添加"C:\Java\jdk1.7.0_01\bin"（根据安装路径填写）。注意：添加部分与前面部分用"；"号隔开，如图 1.16 所示。

（5）JDK 的环境变量已经配置完成，可以通过打开"命令提示符"窗口，输入命令"Java -version"，看到 Java 版本的信息，来确定安装是否成功。首先单击"开始"按钮，然后单击"所有应用程序"→"Windows 系统"→"命令提示符"，如图 1.17 所示。

（6）这时就进入了"命令提示符"窗口。在"命令提示符"窗口中输入"Java -version"。注意：Java 和-version 之间有一个空格，然后按 Enter 键，如图 1.18 所示。

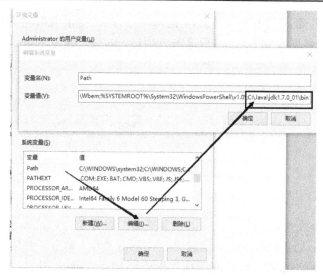

图 1.16 配置 path 变量

图 1.17 Windows 中的命令提示符

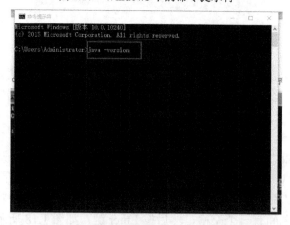

图 1.18 在"命令提示符"窗口中运行"Java-version"命令

(7) JDK 版本信息就全部显示出来了，也表明 JDK 已经安装和配置完成，可以开始进行 Java 开发了，如图 1.19 所示。

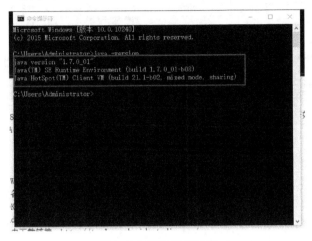

图 1.19 JDK 版本信息

1.3.4 Windows 环境下安装 Android Studio

- 准备工具

(1) JDK 安装包（JDK7 及以上版本）。

(2) Android Studio 安装文件。

国内下载链接：http://tools.android-studio.org/。

官网下载链接：http://developer.android.com/sdk/index.html。

- 安装文件

(1) android-studio-bundle-141.2288178-windows.exe（1.1GB）推荐。

(2) android-studio-ide-141.2288178-windows.exe（348MB）不包含 SDK Tools。

- 说明

(1) 32 位系统和 64 位系统使用同一个安装文件。

(2) 如果计算机中有 Android SDK，可以选择不包含 SDK 的安装版本。

(3) 如果计算机中已经安装过 Android Studio，可以使用压缩文件版本。

(4) 建议使用包含 SDK 的安装文件。

- 安装

这里采用包含 SDK 的安装文件进行讲解，包括了不包含 SDK 的安装文件的安装步骤。如果使用不包含 SDK 的安装文件进行安装，则安装步骤只会比这些步骤少而不会多。如果使用压缩包安装，则可以直接跳过本节内容。

(1) 找到下载的安装文件。

(2) 双击开始安装，如图 1.20 所示。安装解压如图 1.21 所示。

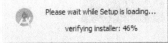

图 1.20 双击开始安装　　　　　　　　　　图 1.21 安装解压

(3) 这里不选择下载 SDK 和模拟器，如图 1.22 所示。
(4) 选择 SDK 的安装路径，如图 1.23 所示。

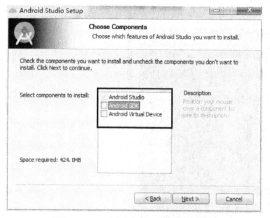

图 1.22　不选择 SDK 和模拟器

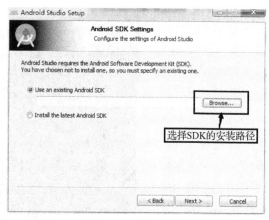

图 1.23　选择安装路径

(5) 设置快捷方式，如图 1.24 所示。
(6) 显示安装进度如图 1.25 所示。

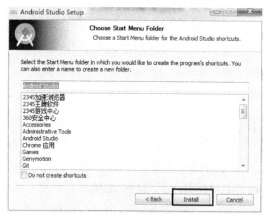

图 1.24　设置快捷方式

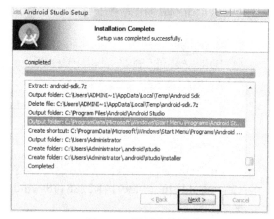

图 1.25　显示安装进度

(7) 安装完成启动 Android Studio，如图 1.26 所示。
(8) Android Studio 启动界面如图 1.27 所示。

图 1.26　安装完成

图 1.27　Android Studio 启动界面

(9) 欢迎界面如图 1.28 所示。

图 1.28　欢迎界面

1.3.5　Mac 环境下安装 Android Studio

- 准备工具

(1) JDK 安装包(JDK7 及以上版本)。

(2) Android Studio 安装文件。

国内下载链接：http://tools.android-studio.org/。

官网下载链接：http://developer.android.com/sdk/index.html。

- 安装文件

android-studio-ide-141.2288178-mac.dmg。

- 安装

(1) 找到下载的安装文件，如图 1.29 所示。

图 1.29　Android Studio 安装文件(Mac OS)

(2) 双击安装，直接将 Android Studio 图标拖入 Application 文件夹即可，如图 1.30 所示。

(3) 在 Launchpad 中找到 Android Studio，如图 1.31 所示。

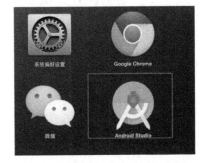

图 1.30　Android Studio 安装过程(Mac OS)　　图 1.31　Launchpad 中的 Android Studio(Mac OS)

(4) 单击"打开"按钮，如图 1.32 所示。

(5) Android Studio 启动界面，如图 1.33 所示。

图1.32　单击"打开"按钮

图1.33　Android Studio 启动界面（Mac OS）

(6) 欢迎界面如图1.34所示。

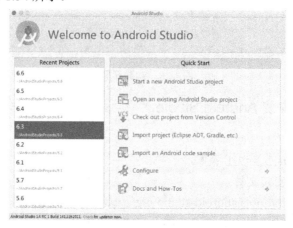

图1.34　欢迎界面（Mac OS）

1.3.6　下载、安装和配置 Android SDK

SDK Tools 其实就是 Android SDK Manager，管理各种版本 SDK 的工具。在 Android SDK 中，包含模拟器、教程、API 文档和示例代码等内容。下面以 Windows 为例详细讲解下载和安装 Android SDK 的步骤。

(1) 打开浏览器，在地址栏中输入 http://www.android-studio.org，进入 Android Studio 中文社区主页，如图1.35所示。

图1.35　Android Studio 中文社区主页

(2) 在主页中找到 SDK TOOLS ONLY R24.3.4（当前最新版本），根据操作系统选择安装包进行下载，强烈建议下载主页推荐版本"installer_r24.3.4-windows.exe（Recommended）"，如图 1.36 所示。

图 1.36　SDK Tools 下载连接

(3) 双击下载的程序，弹出如图 1.37 所示的 SDK Tools 安装向导。
(4) 单击"Next"按钮。如果已经正确安装 JDK，则显示如图 1.38 所示的界面。

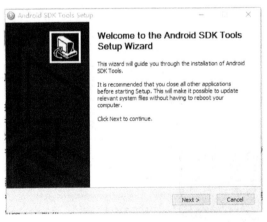

图 1.37　SDK Tools 安装向导　　　　　　图 1.38　配置 JDK 路径

(5) 在图 1.38 中单击"Next"按钮，将提示选择哪种用户可以使用 SDK Tools，这里选择"Install for anyone using this computer"，如图 1.39 所示。
(6) 在图 1.39 中单击"Next"按钮，将显示 Android SDK 安装路径选择窗口。将安装路径修改为"C:\Android\android-sdk"，如图 1.40 所示。
(7) 在图 1.40 中，单击"Next"按钮。此时询问是否在"开始"菜单中创建快捷方式，如图 1.41 所示。单击"Install"按钮开始安装。
(8) 显示安装进度，如图 1.42 所示。

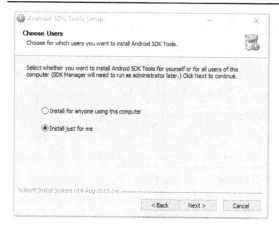

图 1.39　配置用户权限

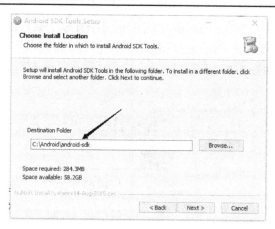

图 1.40　配置安装路径

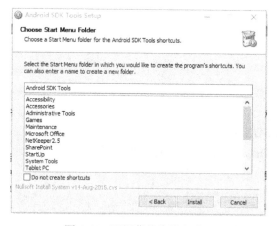

图 1.41　配置菜单快捷方式

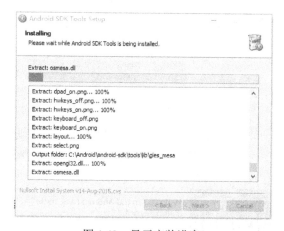

图 1.42　显示安装进度

(9) 安装完成后，弹出如图 1.43 所示窗口，单击 "Next" 按钮。
(10) 在图 1.44 中，单击 "Finish" 按钮。

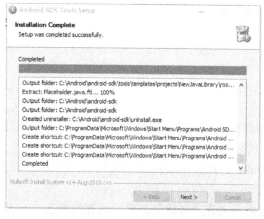

图 1.43　SDK Tools 安装完成（一）

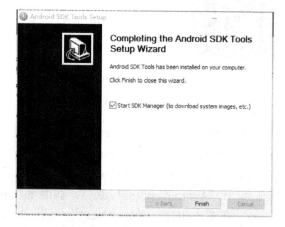

图 1.44　SDK Tools 安装完成（二）

(11) 启动 SDK 管理工具。此时，会自动联网搜索可以下载的 API 等软件包，如图 1.45 所示。
(12) 为了便于今后在不同平台中调试，在搜索完成后选择安装全部软件包，如图 1.46 所示。

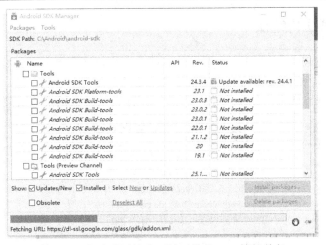

图 1.45 自动联网搜索可以下载的 API 等软件包

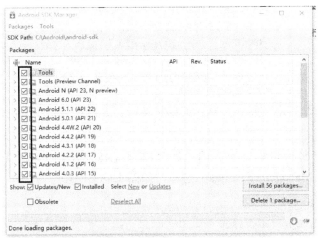

图 1.46 选择安装全部软件包

(13) 在图 1.46 中，单击"Install packages"按钮，安装选中的软件，如图 1.47 所示。

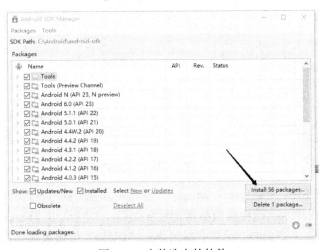

图 1.47 安装选中的软件

(14) 单击"Install"按钮，进行安装。

1.3.7 SDK 下载国内镜像

在国内访问 Google 并不便利，如果 SDK 管理工具自动联网 Google 搜索失败，可以更改国内镜像进行下载更新。

注意：表 1.3 列出的镜像地址可能已失效，请读者自行上网搜索最新镜像地址。

表 1.3　SDK 下载国内镜像地址

大连东软信息学院镜像服务器地址：
http://mirrors.neusoft.edu.cn　端口：80
北京化工大学镜像服务器地址：
IPv4: http://ubuntu.buct.edu.cn/　端口：80
IPv4: http://ubuntu.buct.cn/　端口：80
IPv6: http://ubuntu.buct6.edu.cn/　端口：80
上海 GDG 镜像服务器地址：
http://sdk.gdgshanghai.com　端口：8000
中国科学院开源协会镜像站地址：
IPv4/IPv6: http://mirrors.opencas.cn　端口：80
IPv4/IPv6: http://mirrors.opencas.org　端口：80
IPv4/IPv6: http://mirrors.opencas.ac.cn　端口：80
腾讯镜像服务器地址：
http://android-mirror.bugly.qq.com　端口：8080

● 设置镜像方法

(1) 启动 Android SDK Manager，打开主界面（如图 1.48 所示），选择 "Tools" → "Options"，弹出 "Android SDK Manager - Settings" 窗口，如图 1.49 所示。

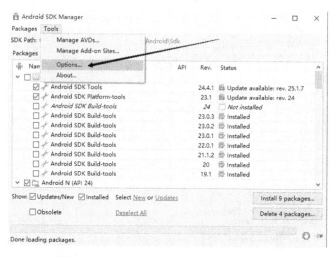

图 1.48　"Android SDK Manager" 主界面

(2) 在 "Android SDK Manager - Settings" 窗口中，在 "HTTP Proxy Server" 和 "HTTP Proxy Port" 输入框内填入上面镜像服务器地址和端口，并且选中 "Force https://... sources to be fetched using http://..." 复选框。设置完成后单击 "Close" 按钮关闭 "Android SDK Manager - Settings" 窗口返回到主界面。

第 1 章　Android 系统与开发环境　　21

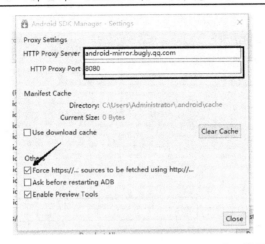

图 1.49　"Android SDK Manager - Settings"窗口设置

（3）依次选择"Packages"、"Reload"，进行重载。

1.3.8　第一次打开 Android Studio 的配置

Android Studio 安装完成后，在第一次启动 Android Studio 前，如果为了避免重新下载新版本的 SDK，应该进行以下操作。

（1）启动 Android Studio 前，为避免第一次打开 Android Studio 时自动重新下载 SDK，应先将 Android Studio 安装目录下的 bin 文件夹中的 idea.properties 文件用记事本打开，在最后增加一行指令：disable.android.first.run=true，如图 1.50 所示。

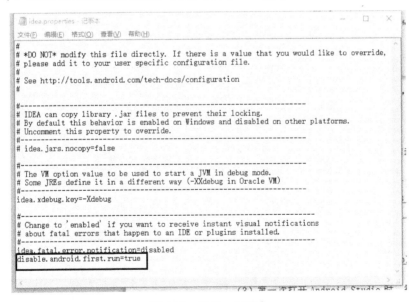

图 1.50　增加一行指令

（2）第一次打开 Android Studio 时，需要配置 JDK 和 SDK，双击 Android Studio 启动图标，打开欢迎对话框，单击"Configure"选项，如图 1.51 所示。

（3）在弹出的配置对话框中，单击"Project Defaults"选项，如图 1.52 所示。

（4）在弹出的项目默认值对话框中单击"Project Structure"选项，如图 1.53 所示。

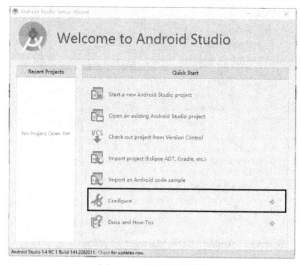

图 1.51　欢迎对话框

图 1.52　配置对话框

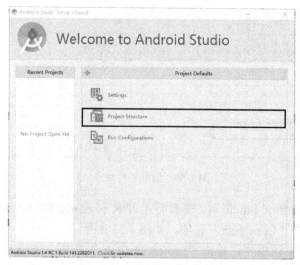

图 1.53　项目默认值对话框

(5) 在弹出的项目结构对话框中指定之前安装的 SDK 和 JDK 的安装路径，完成后单击"Apply"按钮，然后单击"OK"按钮完成配置，如图 1.54 所示。

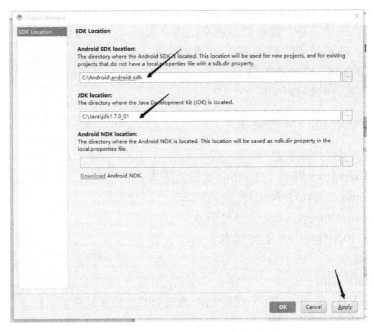

图 1.54 指定 SDK 和 JDK 的安装路径

1.4 Android 模拟器 Genymotion

1.4.1 什么是 Genymotion

由于国内访问 Google 并不便利，无论是离线下载还是在线下载，安装安卓模拟器变得非常困难。Genymotion 工具提供了一套完整的 Android 虚拟环境，迅速成为开发人员、测试人员、推销人员甚至游戏玩家的理想选择。

Genymotion 支持 Windows、Mac、Linux 操作系统，它易于安装、易于运行。它通过功能强大的传感器来测试用户的应用程序，并且完美地融入整个开发环境。

1.4.2 Genymotion 的特性

- 最好的 Android 模拟体验

(1) 支持 OpenGL 加速，提供最好的 3D 性能体验。
(2) 可以从 Google Play 安装应用。
(3) 支持全屏并改善了使用感受。

- 全控制

(1) 可同时启动多个模拟器。
(2) 支持传感器管理，如电池状态、GPS、Accelerator 加速器。
(3) 支持 Shell 控制模拟器。
(4) 完全兼容 ADB，用户可以从主机控制其模拟器。

● 管理设备

(1) 易安装。

(2) 兼容 Microsoft Windows 32/64bit, Mac OSX 10.5+ and Linux 32/64bit。

(3) 可以配置模拟器参数，如屏幕分辨率、内存大小、CPU 数量。

(4) 轻松下载、部署最新的 Genymotion 虚拟设备。

● 从 Android Studio 启动虚拟设备

使用 Genymotion 测试用户的应用。

1.4.3　Genymotion 运行环境要求

● 需要满足以下操作系统之一

(1) Microsoft Windows Vista (32/64bit) 及更高版本。

(2) Linux Ubuntu 12.04 (32/64bit) 及更高版本。

(3) Linux Debian Wheezy 64bit 及更高版本。

(4) Mac OS X 10.8 (64bit) 及更高版本。

● 系统环境要求

(1) 支持 OpenGL 2.0。

(2) CPU 支持 VT-x 或者 AMD-V 虚拟化，通过 BIOS 设置开启。

(3) 至少 2GB 内存。

(4) 至少提供 400MB 的硬盘空间供 Genymotion 安装，至少 2GB 硬盘空间用来部署 Genymotion 虚拟设备。有些可能需要 8GB 以上，这主要取决于用户安装在虚拟机里的应用程序。

(5) 可用网络连接(用于安装和更新)。

(6) 系统显示分辨率不低于 1024×768。

● 应用程序要求

Oracle VirtualBox 4.1 或以上(版本越高越好)。

1.4.4　注册和下载 Genymotion 模拟器

(1) 注册 Genymotion 账号：打开浏览器，在地址栏中输入 "http://www.genymotion.com"，进入如图 1.55 所示的 Genymotion 官网主页，单击右上角的 "Sign in" 按钮进入登录界面。

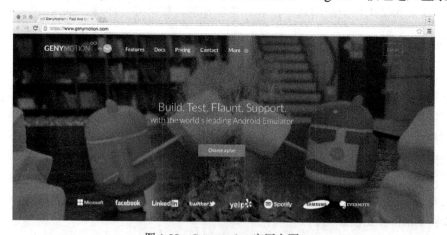

图 1.55　Genymotion 官网主页

(2) 单击"Account an creation"按钮创建个人账号，依次填写用户名、电子邮件、密码、选择公司规模(可选项)、用途类型(可选项)，勾选最后一项"I accept terms of the privacy statement"(同意条款)，如图 1.56 所示。

(3) 注册成功后，将弹出如图 1.57 所示界面，显示了用户注册的 Genymotion 账号(用户注册时使用的 E-mail 地址)。同时，注册成功后，系统会发送一封邮件到用户的邮箱，请一定要打开邮件完成验证。

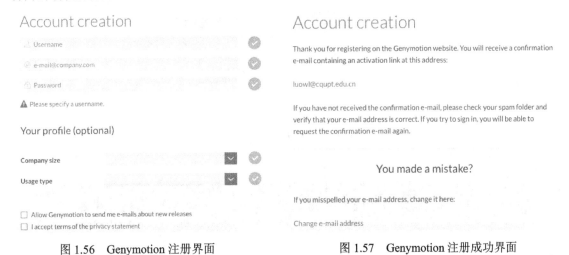

图 1.56　Genymotion 注册界面　　　　图 1.57　Genymotion 注册成功界面

(4) 回到主页(如图 1.58 所示)，单击屏幕中间的"Choose a plan"按钮，进入下载用户类型选择界面，如图 1.59 所示。

图 1.58　Genymotion 官网主页

(5) 在界面中，单击"Individual"按钮，进入个人下载界面，如图 1.60 所示。

(6) Genymotion 对于个人学习使用是免费的，对于专业个人和对于 2 人以上的小型公司将收取 136 美元 1 年的费用。单击"BASIC"栏下的"Get started"按钮，进行免费下载。

(7) 在如图 1.61 所示的下载界面中根据个人计算机硬件和操作系统类型选择合适的安装文件。下面以 Windows 系统为例进行讲解。对于计算机没有安装过 Virtualbox 的用户，建议下载 with VirtualBox 版本，在图 1.61 中单击"Download for Windows"按钮完成下载。

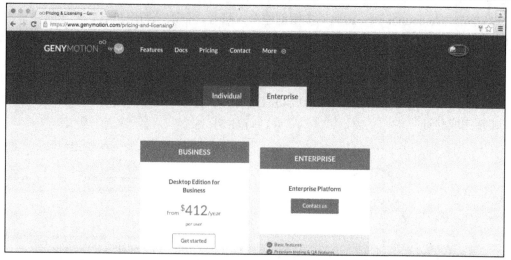

图 1.59 下载用户类型选择界面

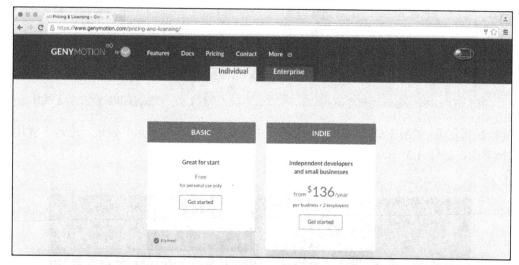

图 1.60 个人下载界面

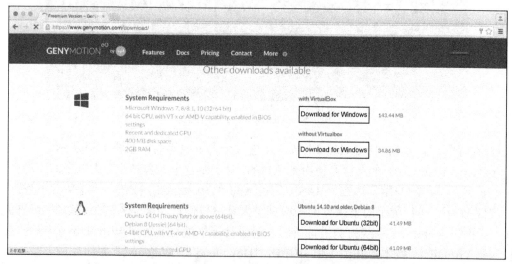

图 1.61 根据个人计算机硬件和操作系统类型选择合适的安装文件

1.4.5 安装 Genymotion 模拟器

(1) 双击运行下载的 Genymotion 安装文件，选择中文语言并单击"Next"按钮，如图 1.62 所示。

(2) 可更改安装路径，单击"Browse"按钮，软件默认的路径为"C:\Program Files\Genymobile\Genymotion"，然后单击"Next"按钮，如图 1.63 所示。

图 1.62　Genymotion 安装向导　　　　图 1.63　Genymotion 安装路径对话框

(3) 在弹出的对话框中选择是否创建快捷菜单"Don't create a Start Menu folder"，然后单击"Next"按钮，如图 1.64 所示。

(4) 在弹出的对话框中选择是否创建桌面快捷方式（Create a desktop icon），单击"Next"→"Install"→"Finish"按钮，如图 1.65 所示。

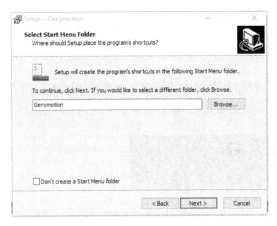
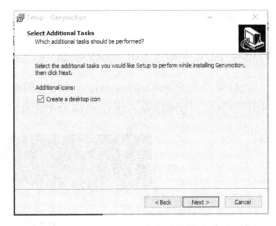

图 1.64　Genymotion 创建快捷菜单对话框　　　　图 1.65　Genymotion 创建桌面快捷方式对话框

(5) 在安装完 Genymotion 后，会继续安装 VirtualBox。在 VirtualBox 安装向导对话框中，单击"Next"按钮，如图 1.66 所示。

(6) 单击"Browse"按钮以更改 Location 的地址，VirtualBox 软件默认路径为"C:\Program Files\Oracle\VirtualBox\"，然后单击"Next"按钮，如图 1.67 所示。

(7) 询问是否现在安装，单击"Yes"按钮，如图 1.68 所示。

(8) 单击"Install"按钮开始安装（如图 1.69 所示），然后单击"Finish"按钮完成安装。

图 1.66　VirtualBox 安装向导对话框

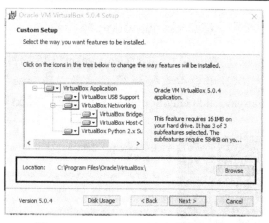

图 1.67　VirtualBox 更改安装路径对话框

图 1.68　VirtualBox 确认安装对话框

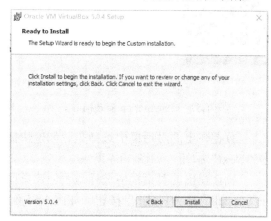

图 1.69　开始安装 Genymotion 对话框

1.4.6　使用 Genymotion 模拟器

（1）第一次进入 Genymotion，系统会检查你是否拥有安卓虚拟设备。如果没有安装，则会弹出对话框，询问你是否现在添加一个虚拟设备，如图 1.70 所示单击"Yes"按钮就可以了。

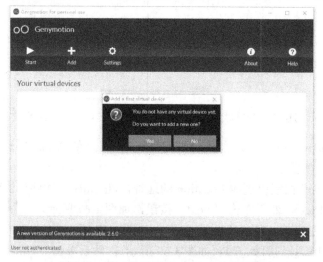

图 1.70　第一次启动 Genymotion 对话框

(2)创建一个新的虚拟设备"Create a new virtual device"。需要输入用户名和密码验证。例如：qq 邮箱*******@qq.com（注意：如果验证不通过，请到邮箱确认是否已经验证过），如图 1.71 所示。

图 1.71　Genymotion 登录界面

(3)验证成功后，可以看到有很多虚拟设备，如 Samsung Galaxy S3、S4 等，如图 1.72 所示。选择想添加的虚拟设备，选择后单击"Next"按钮。

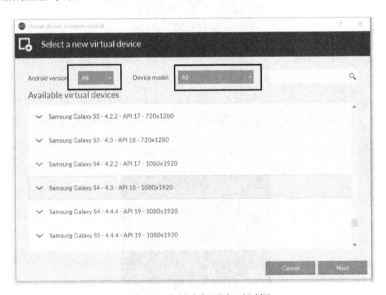

图 1.72　选择虚拟设备对话框

(4)下载安装，在图 1.73 中，下载到 100%时单击"Finish"按钮。

(5)回到主界面，选择一个已经添加的模拟器，单击"Start（启动）"按钮启动模拟器，如图 1.74 所示。

(6)启动模拟器，将启动如图 1.75 所示的手机模拟界面。

Android 应用程序开发教程——Android Studio 版

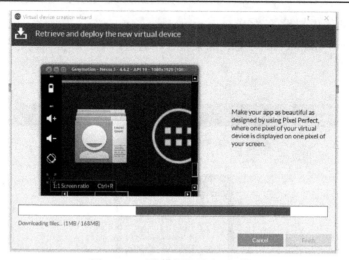

图 1.73 下载模拟器进度对话框

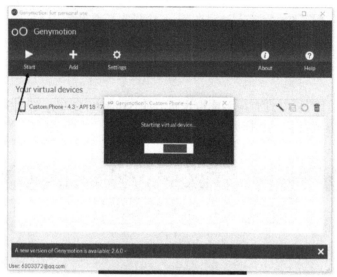

图 1.74 启动模拟器

图 1.75 手机模拟界面

1.4.7 给 Android Studio 安装 Genymotion 插件

（1）打开 Android Studio，依次选择菜单项"File"→"Settings"，如图 1.76 所示。

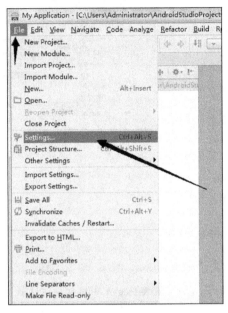

图 1.76 Android Studio "File（文件）"菜单

（2）在打开的对话框中找到"Plugins"设置项，单击下端的"Browse repositories"选项，如图 1.77 所示。

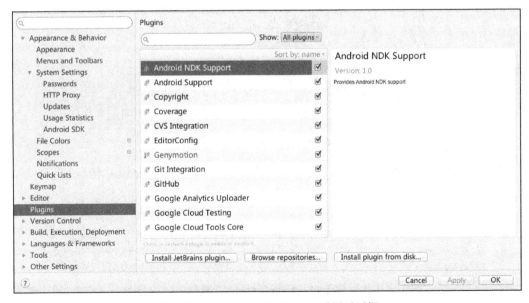

图 1.77 Android Studio Plugins 设置对话框

（3）在搜索栏中输入 Genymotion 关键字，自动在右边的框中显示已经搜索到的插件，单击"Install plugin"按钮安装，如图 1.78 所示。

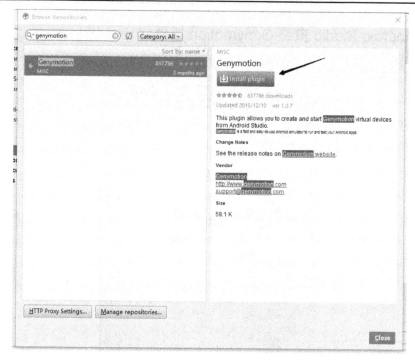

图 1.78　安装 Genymotion 插件对话框

（4）单击"Yes"按钮开始下载，如图 1.79 所示。

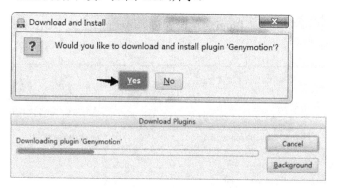

图 1.79　Genymotion 下载界面

（5）安装后重启 Android Studio，在 Android Studio 工具栏中看到新增一个图标，如图 1.80 所示。

图 1.80　Android Studio 工具栏

（6）初次单击图标还需要设置 Genymotion 安装目录，如图 1.81 所示。

（7）设置好安装目录后，再次单击工具栏上的图标后进行模拟器的设置和启动，选中下载的模拟器后单击"Start"按钮以启动模拟器，如图 1.82 所示。

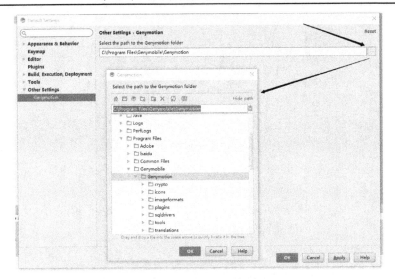

图 1.81　设置 Genymotion 安装目录对话框

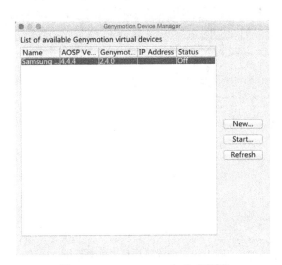

图 1.82　Genymotion 启动对话框

1.5　新建一个 Hello World 项目

配置好 JDK 和 Android SDK 后，就可以开始新建 Android 项目了，具体步骤如下。

(1) 启动 Android Studio，打开欢迎对话框，单击"Start a new Android Studio project"选项，如图 1.83 所示。

(2) 在"Application Name"文本框中输入 HelloWorld，在"Company Domain"文本框中输入 administrator.example.com（根据自己需要改为姓名拼音.com 或公司英文名称.com），在"Project location"文本框中选择项目保存路径，然后单击"Next"按钮，如图 1.84 所示。

(3) 在弹出的如图 1.85 所示的对话框中选择"Phone and Tablet"选项，在"Minimum SDK"选项中选择运行 Android 最低 SDK 版本要求（此处以 API：18 Android 4.3（Jelly Bean）为例），然后单击"Next"按钮。

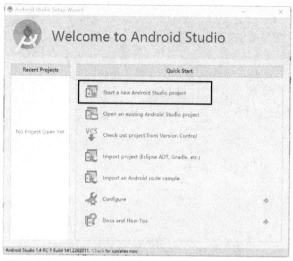

图 1.83　在欢迎对话框中新建一个 Android Studio 项目

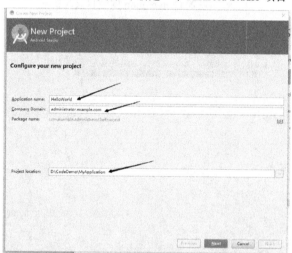

图 1.84　设置新项目名称对话框

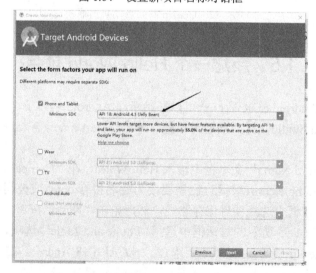

图 1.85　设置新项目最低 SDK 版本对话框

(4)在弹出的对话框中选择"Empty Activity"选项,然后单击"Next"按钮,如图 1.86 所示。

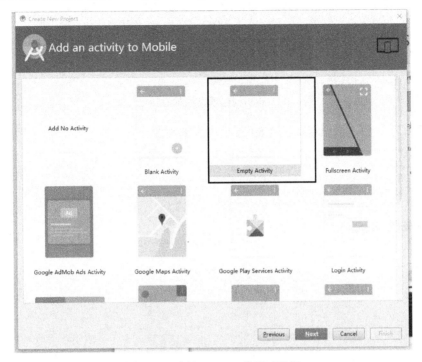

图 1.86　选择 Activity 类型对话框

(5)在弹出的对话框中单击"Finish"按钮(该对话框将设置 Activity、Layout 的名称,建议初学者不要修改此对话框的默认设置),如图 1.87 所示。

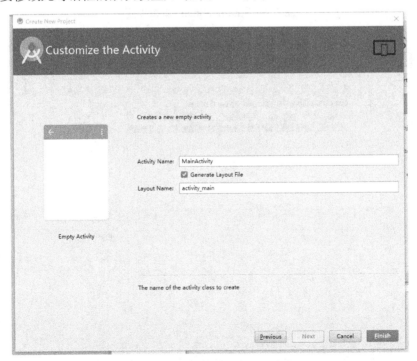

图 1.87　设置 Activity 名称对话框

（6）加载新创建的 Android Studio 项目，会出现如图 1.88 所示的对话框。

图 1.88 创建项目加载对话框

（7）创建好项目后，会出现如图 1.89 所示的编辑界面。

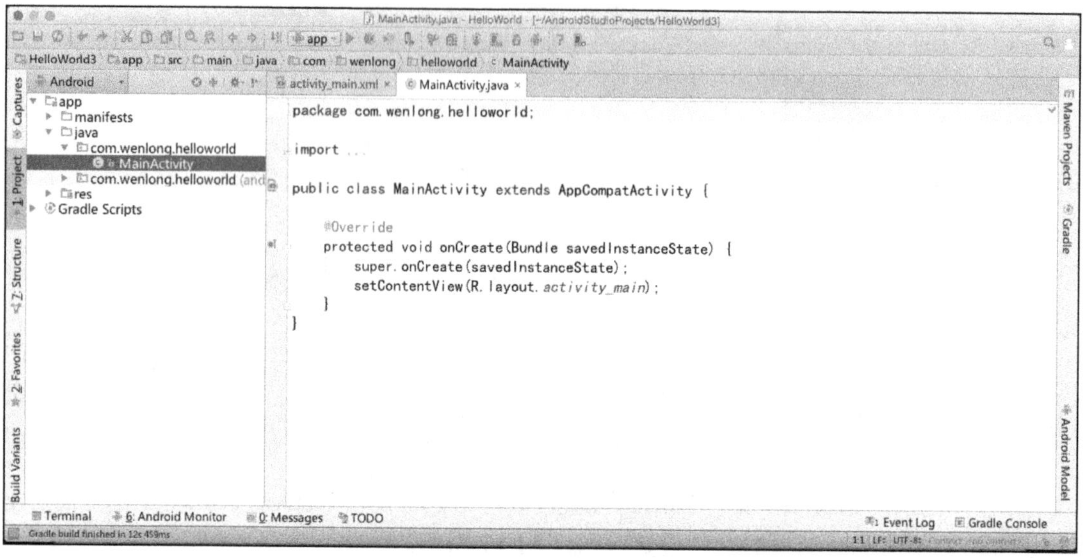

图 1.89 Android Studio 编辑界面

（8）启动 Genymotion 模拟器：单击 Android Studio 工具条中的 按钮，弹出如图 1.90 所示的对话框，选择一个已经下载的模拟器平台，然后单击"Start"按钮。

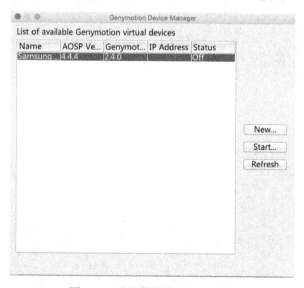

图 1.90 选择模拟器平台对话框

(9)模拟器平台启动成功后,弹出如图 1.91 所示的手机模拟界面,滑动屏幕锁以解锁屏幕,如图 1.92 所示。

图 1.91 手机模拟界面(一)

图 1.92 手机模拟界面(二)

(10)单击 Android Studio 工具条中的 ▶ 按钮,弹出如图 1.93 所示的项目运行方式选择对话框。确认有刚才运行的 Genymotion 模拟器后,单击"OK"按钮。

(11)在模拟器中查看 HelloWorld 项目的运行效果,如图 1.94 所示。

图 1.93 Android Studio 项目运行方式选择对话框

图 1.94 HelloWorld 项目的运行效果

第 2 章 Android UI 设计

2.1 Android UI 布局

Android UI 布局一般使用布局管理器、ListView（列表视图）和 GridView（网格视图）三种，下面对它们分别进行详细介绍。

Android 布局管理器可以很方便地控制各个组件的位置和大小，同时根据屏幕大小，管理容器内的控件，自动适配组件在手机屏幕中的位置。Android 中提供了如图 2.1 所示的 7 种布局管理器：FrameLayout（帧布局管理器）、LinearLayout (Horizontal)（水平线性布局管理器）、LinearLayout（Vertical）（垂直线性布局管理器）、TableLayout（表格布局管理器）、TableRow（表格行布局管理器）、GridLayout（网格布局管理器）、RelativeLayout（相对布局管理器）。这 7 种布局管理器的布局如图 2.2 所示。

图 2.1 Android Studio 中的 7 种布局管理器

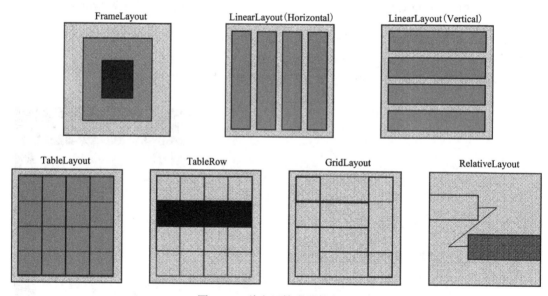

图 2.2 7 种布局管理器的布局

在一个 Android 应用程序中，用户界面通过 View 和 ViewGroup 对象构建。Android 中有很多种 View 和 ViewGroup，它们都继承自 View 类。View 对象是 Android 平台上表示用户界面的基本单元。布局方式是指一组 View 元素如何布局，准确地说是一个 ViewGroup 中包含的一些 View 如何布局。这里介绍的 View 的布局方式的类都直接或间接继承自 ViewGroup 类，如图 2.3 所示。

第 2 章 Android UI 设计

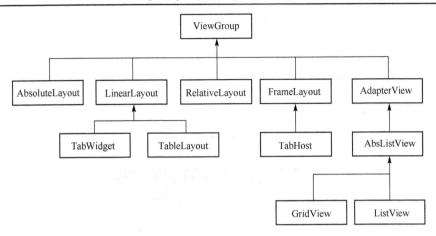

图 2.3 Android Layouts

其实，所有的布局方式都可以归类为 ViewGroup 的 5 个类别，即 ViewGroup 的 5 个直接子类。其他的一些布局都扩展自这 5 个类。

【例 2.1】 FrameLayout（帧布局管理器）实例

这个例子将使用 FrameLayout（帧布局管理器）创建一个简单的 Android 应用，下面将通过表 2.1 中的步骤来实现。

表 2.1 FrameLayout（帧布局管理器）实例步骤

步骤	描述
1	打开 Android Studio 创建一个 Android 应用，Application name（应用名）取名为 FrameLayout，Company Domain（公司域）取名为 Example.com
2	Minimum SDK 选择 API 18: Android 4.3（Jelly Bean）
3	选择 Empty Activity
4	不需要修改 Activity Name，使用默认值，单击"Finish"按钮（完成）
5	启动 Genymotion 模拟器，然后在 Android 工程中做如下代码修改

将工程 res/mipmap 目录下的 ic_launcher.png 图片文件复制、粘贴到 res/drawable 目录下，并按如下代码修改 res/layout 目录下的布局文件 activity_main.xml。

```xml
<?xml version="1.0" encoding="utf-8"?>
<FrameLayout xmlns:android="http://schemas.android.com/apk/res/android"
    android:layout_width="fill_parent"
    android:layout_height="fill_parent">
<!--添加居中显示的安卓图标图片 ImageView，将显示在最下层-->
    <ImageView
        android:src="@drawable/th"
        android:scaleType="fitCenter"
        android:layout_height="392dp"
        android:layout_width="match_parent"
        android:layout_gravity="center_vertical" />
<!--添加居中显示的"Frame Demo"文字 TextView，将显示在最上层-->
    <TextView
        android:text="Frame Demo"
        android:textSize="30px"
```

```
        android:textStyle="bold"
        android:layout_height="fill_parent"
        android:layout_width="fill_parent"
        android:gravity="center"
        android:layout_gravity="left|bottom" />
</FrameLayout>
```

在 Genymotion 中运行程序的效果如图 2.4 所示。

图 2.4　FrameLayout(帧布局管理器)实例运行效果

【例 2.2】 LinearLayout(Horizontal/ Vertical)(水平/垂直线性布局管理器)实例

这个例子将使用 LinearLayout(Horizontal/Vertical)(水平/垂直线性布局管理器)创建一个简单的 Android 应用,下面将通过表 2.2 中的步骤来实现。

表 2.2　LinearLayout(Horizontal/ Vertical)(水平/垂直线性布局管理器)实例步骤

步骤	描述
1	打开 Android Studio 创建一个 Android 应用,Application name(应用名)取名为 LinearLayout,Company Domain(公司域)取名为 Example.com
2	Minimum SDK 选择 API 18: Android 4.3 (Jelly Bean)
3	选择 Empty Activity
4	不需要修改 Activity Name,使用默认值,单击"Finish"按钮(完成)
5	启动 Genymotion 模拟器,然后在 Android 工程中做如下代码修改

按如下代码修改工程的 res/layout 目录下的布局文件 activity_main.xml:

```
<?xml version="1.0" encoding="utf-8"?>
<LinearLayout xmlns:android="http://schemas.android.com/apk/res/android"
    android:layout_width="fill_parent"
```

```xml
    android:layout_height="fill_parent"
    android:orientation="vertical" >
<!--添加 start_service 按钮-->
<Button android:id="@+id/btnStartService"
    android:layout_width="142dp"
    android:layout_height="wrap_content"
    android:text="start_service"/>
<!--添加 pause_service 按钮-->
<Button android:id="@+id/btnPauseService"
    android:layout_width="142dp"
    android:layout_height="wrap_content"
    android:text="pause_service"/>
<!--添加 stop_service 按钮-->
<Button android:id="@+id/btnStopService"
    android:layout_width="142dp"
    android:layout_height="wrap_content"
    android:text="stop_service"/>
</LinearLayout>
```

在 Genymotion 中运行程序的效果如图 2.5 所示。

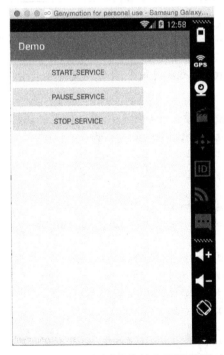

图 2.5　LinearLayout(Vertical)(垂直线性布局管理器)实例运行效果

将 res/layout 目录下的布局文件 activity_main.xml 代码中的 android:orientation= "vertical" 改为 android:orientation="horizontal"再运行一次这个项目，在 Genymotion 中运行程序，单击模拟器的◇按钮就使手机屏幕转为横向，运行效果如图 2.6 所示。

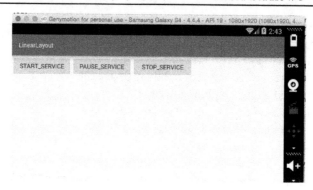

图 2.6　LinearLayout（Horizontal）（水平线性布局管理器）实例运行效果

【例 2.3】 TableLayout（表格布局管理器）、TableRow（表格行布局管理器）实例

这个例子将使用 TableLayout（表格布局管理器）、TableRow（表格行布局管理器）创建一个简单的 Android 应用，下面将通过表 2.3 中的步骤来实现。

表 2.3　TableLayout（表格布局管理器）、TableRow（表格行布局管理器）实例步骤

步骤	描述
1	打开 Android Studio 创建一个 Android 应用，Application name（应用名）取名为 TableLayout，Company Domain（公司域）取名为 Example.com
2	Minimum SDK 选择 API 18: Android 4.3（Jelly Bean）
3	选择 Empty Activity
4	不需要修改 Activity Name，使用默认值，单击"Finish"按钮（完成）
5	启动 Genymotion 模拟器，然后在 Android 工程中做如下代码修改

按如下代码修改工程的 res/layout 目录下的布局文件 activity_main.xml：

```xml
<?xml version="1.0" encoding="utf-8"?>
<TableLayout xmlns:android="http://schemas.android.com/apk/res/android"
    android:layout_width="fill_parent"
    android:layout_height="fill_parent">

    <TableRow
        android:layout_width="fill_parent"
        android:layout_height="fill_parent">
    <TextView
        android:text="Time"
        android:layout_width="wrap_content"
        android:layout_height="wrap_content"
        android:layout_column="1" />
    <TextClock
        android:layout_width="wrap_content"
        android:layout_height="wrap_content"
        android:id="@+id/textClock"
        android:layout_column="2" />
    </TableRow>

    <TableRow
```

```xml
        <TextView
            android:text="First Name"
            android:layout_width="wrap_content"
            android:layout_height="wrap_content"
            android:layout_column="1" />
        <EditText
            android:width="200px"
            android:layout_width="wrap_content"
            android:layout_height="wrap_content" />
    </TableRow>

    <TableRow>
        <TextView
            android:text="Last Name"
            android:layout_width="wrap_content"
            android:layout_height="wrap_content"
            android:layout_column="1" />
        <EditText
            android:width="100px"
            android:layout_width="wrap_content"
            android:layout_height="wrap_content" />
    </TableRow>

    <TableRow
        android:layout_width="fill_parent"
        android:layout_height="fill_parent">
        <RatingBar
            android:layout_width="wrap_content"
            android:layout_height="wrap_content"
            android:id="@+id/ratingBar"
            android:layout_column="2" />
    </TableRow>

    <TableRow
        android:layout_width="fill_parent"
        android:layout_height="fill_parent"/>
    <TableRow
        android:layout_width="fill_parent"
        android:layout_height="fill_parent">
        <Button
            android:layout_width="wrap_content"
            android:layout_height="wrap_content"
            android:text="Submit"
            android:id="@+id/button"
            android:layout_column="2" />
    </TableRow>
</TableLayout>
```

在 Genymotion 中运行程序的效果如图 2.7 所示。

图 2.7 TableLayout（表格布局管理器）、TableRow（表格行布局管理器）实例运行效果

【例 2.4】 RelativeLayout（相对布局管理器）实例

这个例子将使用 RelativeLayout（相对布局管理器）创建一个简单的 Android 应用，下面将通过表 2.4 中的步骤来实现。

表 2.4 RelativeLayout（相对布局管理器）实例步骤

步骤	描述
1	打开 Android Studio 创建一个 Android 应用，Application name（应用名）取名为 RelativeLayout，Company Domain（公司域）取名为 Example.com
2	Minimum SDK 选择 API 18: Android 4.3（Jelly Bean）
3	选择 Empty Activity
4	不需要修改 Activity Name，使用默认值，单击"Finish"按钮（完成）
5	启动 Genymotion 模拟器，然后在 Android 工程中做如下代码修改

按如下代码修改工程的 res/layout 目录下的布局文件 activity_main.xml：

```xml
<?xml version="1.0" encoding="utf-8"?>
<RelativeLayout xmlns:android="http://schemas.android.com/apk/res/android"
    android:layout_width="fill_parent"
    android:layout_height="fill_parent"
    android:paddingLeft="16dp"
    android:paddingRight="16dp" >


    <EditText
        android:id="@+id/name"
```

```xml
        android:layout_width="fill_parent"
        android:layout_height="wrap_content"
        android:hint="Enter your name" />

    <LinearLayout
        android:orientation="vertical"
        android:layout_width="fill_parent"
        android:layout_height="fill_parent"
        android:layout_alignParentStart="true"
        android:layout_below="@+id/name">

        <Button
            android:layout_width="wrap_content"
            android:layout_height="wrap_content"
            android:text="New Button"
            android:id="@+id/button" />

        <Button
            android:layout_width="wrap_content"
            android:layout_height="wrap_content"
            android:text="New Button"
            android:id="@+id/button2" />
    </LinearLayout>
</RelativeLayout>
```

在 Genymotion 中运行程序的效果如图 2.8 所示。

图 2.8　RelativeLayout(相对布局管理器)实例运行效果

2.2 ListView（列表视图）

ListView（列表视图）是 Android 中常用的 UI 组件之一，它将以垂直列表的形式列出要显示的列表项目。一般来说，ListView 都是和 Adapter（适配器）一起配合使用，数组数据或者数据库数据都将通过 Adapter 把值传递给 ListView 组件。图 2.9 中显示出 ListView 运行后的效果图。

图 2.9　ListView（列表视图）实例运行效果

Adapter 实际上是 UI 组件和数据源之间的一座桥梁，Adapter 将数据从数据源中得到后传递给 AdapterView（适配器视图），AdapterView 则将数据呈现在 Spinner（列表选择框）、ListView（列表视图）、GrideView（网格视图）等 UI 组件中。实际上，ListView、GrideView 是 AdapterView 的子类，它们通常和一个 Adapter 在一起使用，Adapter 负责收集外部数据，然后再把数据项以 View 的形式显示在界面上。

Android 提供了几种类型的 Adapter，它们都是 Adapter 的子类，常用的 Adapter 类型有 ArrayAdapter（数组适配器）、BaseAdapter（基本适配器）、CursorAdapter（游标适配器）、SimpleCursorAdapter（简单游标适配器）、SpinnerAdapter（列表适配器）和 WrapperListAdapter（封装列表适配器）。不同的适配器将用于绑定不同类型的数据，比如 ArrayAdapter 就适合绑定数组数据。

2.2.1 ListView（列表视图）的常用属性

ListView（列表视图）的常用属性如表 2.5 所示。关于该控件的其他属性，可以参阅 Android 官方提供的完整的 API 文档。

表 2.5 ListView 的常用属性

属性	描述
android:id	用于设置 ListView 的名称
android:divider	用于为 ListView 设置分割条,既可以用颜色分割,也可以用 Drawable 资源分割
android:dividerHeight	用于设置分割条的高度
android:entries	用于通过数组资源为 ListView 指定列表项
android:footerDividersEnabled	用于设置是否在 footer view 之前绘制分割条,默认为 true,设置为 false 时,表示不绘制。需要通过 ListView(列表视图)提供的 addFooterView()方法为 ListView(列表视图)添加 footer view
android:headerDividersEnabled	用于设置是否在 header view 之前绘制分割条,默认为 true,设置为 false 时,表示不绘制。需要通过 ListView(列表视图)提供的 addHeaderView()方法为 ListView(列表视图)添加 header view

2.2.2 ArrayAdapter(数组适配器)

当需要为一个控件赋值,而赋值的数据源是一个数组时,就应该考虑使用 ArrayAdapter(数组适配器)。

首先创建一个适配器对象,然后使用 ArrayAdapter 类的构造方法 ArrayAdapter(Context context, int textViewResourceId, T[] objects)实例化一个 ArrayAdapter 对象。具体代码如下:

```
string[] StringArray=new String[]{"学生","老师","工作人员",}
ArrayAdapter adapter = new
ArrayAdapter<String>(this,R.layout.ListView,StringArray);
```

当创建好一个 ArrayAdapter 后,就可以简单地通过 ListView 的 setAdapter()方法进行调用,代码如下:

```
ListView listView = (ListView) findViewById(R.id.listview);
listView.setAdapter(adapter);
```

【例 2.5】 ListView(列表视图)实例

这个例子将使用 ListView(列表视图)控件来创建一个简单的 Android 应用,下面将通过表 2.6 中的步骤来实现。

表 2.6 ListView(列表视图)实例步骤

步骤	描述
1	打开 Android Studio 创建一个 Android 应用,Application name(应用名)取名为 ListDisplay,Company Domain(公司域)取名为 Example.com
2	Minimum SDK 选择 API 18: Android 4.3 (Jelly Bean)
3	选择 Empty Activity
4	不需要修改 Activity Name,使用默认值,单击"Finish"按钮(完成)
5	在工程中找到 ras/layout 目录中的 activity_main.xml 文件,在其中添加一个 ListView(列表视图)控件
6	在 ras/layout 目录下添加一个名为 activity_listview.xml 的文件。这个 xml 文件将用于显示所有的列表项,可以用这个文件对字体、间距、颜色等进行自定义
7	启动 Genymotion 模拟器,然后在 Android 工程中做如下代码修改

修改工程的 res/layout 目录下的布局文件 activity_main.xml,修改代码如下:

```
<?xml version="1.0" encoding="utf-8"?>
<LinearLayout xmlns:android="http://schemas.android.com/apk/res/android"
```

```xml
    xmlns:tools="http://schemas.android.com/tools"
    android:layout_width="match_parent"
    android:layout_height="match_parent"
    android:orientation="vertical"
    tools:context=".ListActivity" >

    <ListView
        android:id="@+id/mobile_list"
        android:layout_width="match_parent"
        android:layout_height="wrap_content" >
    </ListView>

</LinearLayout>
```

修改新建工程的 res/layout 目录下的布局文件 activity_listview.xml，将默认添加的布局代码删除，然后添加如下代码：

```xml
<?xml version="1.0" encoding="utf-8"?>

<!-- Single List Item Design -->
<TextView xmlns:android="http://schemas.android.com/apk/res/android"
    android:id="@+id/label"
    android:layout_width="fill_parent"
    android:layout_height="fill_parent"
    android:padding="10dip"
    android:textSize="16dip"
    android:textStyle="bold" >
</TextView>
```

修改新建工程的 java/com.example.listdisplay 目录下的 java 文件 MainActivity.java，将默认添加的布局代码删除，然后添加如下代码：

```java
package com.example.listdisplay;
import android.os.Bundle;
import android.support.v7.app.AppCompatActivity;
import android.widget.ArrayAdapter;
import android.widget.ListView;

public class MainActivity extends AppCompatActivity {
    // Array of strings...
    String[] mobileArray={"Android","IPhone","WindowsMobile","Blackberry",
                          "WebOS","Ubuntu","Windows7","Max OS X"};
    @Override
    protected void onCreate(Bundle savedInstanceState) {
        super.onCreate(savedInstanceState);
        setContentView(R.layout.activity_main);
        ArrayAdapter adapter = new ArrayAdapter<String>(this, R.layout.
                        activity_listview, mobileArray);

        ListView listView = (ListView) findViewById(R.id.mobile_list);
        listView.setAdapter(adapter);

    }
}
```

在 Genymotion 中运行程序的效果如图 2.10 所示。

图 2.10　ListView(列表视图)实例运行效果

2.3　GridView(网格视图)

GridView(网格视图)也是 Android 中常用的 UI 组件之一，它是按照行、列分布的方式来显示多个组件的，通常用于显示图片或图标等。图 2.11 中显示出 GridView 运行后的效果。

图 2.11　GridView(网格视图)实例运行效果

GridView（网格视图）的常用属性如表 2.7 所示。关于该控件的其他属性，可以参阅 Android 官方提供的完整的 API 文档。

表 2.7 GridView 的常用属性

属性	描述
android:id	用于设置 GridView 的名称
android:columnWidth	用于设置列的宽度
android:gravity	用于设置对齐方式
android:horizontalSpacing	用于设置各元素之间的水平间距
android:numColumns	用于设置列数，其属性值通常大于 1，如果值较大，如 100 或者为 auto_fit，则将在可用空间中显示尽量多的列数
android:stretchMode	用于设置拉伸模式，其中属性可以为： ● none：不拉伸 ● spacingWidth：仅拉伸元素之间的间距 ● columnWidth：仅拉伸表格元素本身 ● spacingWidthUniform：表格元素本身、元素之间的间距一起拉伸
android:verticalSpacing	用于设置各元素之间的垂直间距

GridView 和 ListView 类似，都需要通过 Adapter 来提供要显示的数据。在使用 GridView 时，通常使用 ImageAdapter（图片适配器）为 GridView 组件提供数据。下面通过一个具体的实例演示如何通过 ImageAdapter 指定内容的方式创建 GridView。

【例 2.6】 GridView（网格视图）实例

这个例子将使用 GridView（网格视图）控件来创建一个简单的 Android 应用，下面将通过表 2.8 中的步骤来实现。

注意：Android 虚拟机不允许单个程序中位图超过 8MB 的内存。当大量图片加载时很容易出现内存溢出报错，而且如果位图突然要占用大量的内存，即使和之前已经剩下的内存加起来不到 8MB，系统也会报错，原因是它变"敏感"了。处理方法有将图片压缩大小、及时回收内存、使用缓存等。在例 2.6 中可以通过简单的减小图片大小再将其复制到工程中来避免程序报错。

表 2.8 GridView（网格视图）实例步骤

步骤	描述
1	打开 Android Studio 创建一个 Android 应用，Application name（应用名）取名为 GridDisplay，Company Domain（公司域）取名为 Example.com
2	Minimum SDK 选择 API 18: Android 4.3 (Jelly Bean)
3	选择 Empty Activity
4	不需要修改 Activity Name，使用默认值，单击"Finish"按钮（完成）
5	在工程中找到 ras/layout 目录中的 activity_main.xml 文件，在其中添加一个 GridView（网格视图）控件
6	将图片 sample0.jpg、sample1.jpg、sample2.jpg、sample3.jpg、sample4.jpg、sample5.jpg、sample6.jpg 和 sample7.jpg 复制、粘贴到工程目录 res/drawable 目录下
7	在 java/com.example.griddisplay 目录下创建一个继承自 BaseAdapter（基本适配器）类的 ImageAdapter（图片适配器）子类，取名为 ImageAdapter.java，用于为 GridView（网格视图）控件提供图片数据
8	启动 Genymotion 模拟器，然后在 Android 工程中做如下代码修改

修改工程 res/layout 目录下的布局文件 activity_main.xml，修改代码如下：

```xml
<?xml version="1.0" encoding="utf-8"?>
<GridView xmlns:android="http://schemas.android.com/apk/res/android"
    android:id="@+id/gridview"
    android:layout_width="fill_parent"
    android:layout_height="fill_parent"
    android:columnWidth="90dp"
    android:numColumns="auto_fit"
    android:verticalSpacing="10dp"
    android:horizontalSpacing="10dp"
    android:stretchMode="columnWidth"
    android:gravity="center"
/>
```

修改工程的java/com.example.griddisplay 目录下java 文件ImageAdapter.java，修改代码如下：

```java
package com.example.griddisplay;

import android.content.Context;
import android.view.View;
import android.view.ViewGroup;
import android.widget.BaseAdapter;
import android.widget.GridView;
import android.widget.ImageView;

public class ImageAdapter extends BaseAdapter {
    private Context mContext;

    // Constructor
    public ImageAdapter(Context c) {
        mContext = c;
    }

    public int getCount() {
        return mThumbIds.length;
    }

    public Object getItem(int position) {
        return null;
    }

    public long getItemId(int position) {
        return 0;
    }

    // create a new ImageView for each item referenced by the Adapter
    public View getView(int position, View convertView, ViewGroup parent) {
        ImageView imageView;

        if (convertView == null) {
            imageView = new ImageView(mContext);
```

```
        imageView.setLayoutParams(new GridView.LayoutParams(85, 85));
        imageView.setScaleType(ImageView.ScaleType.CENTER_CROP);
        imageView.setPadding(8, 8, 8, 8);
    }
    else
    {
        imageView = (ImageView) convertView;
    }
    imageView.setImageResource(mThumbIds[position]);
    return imageView;
}

// Keep all Images in array
public Integer[] mThumbIds = {
        R.drawable.sample2, R.drawable.sample3,
        R.drawable.sample4, R.drawable.sample5,
        R.drawable.sample6, R.drawable.sample7,
        R.drawable.sample0, R.drawable.sample1,
        R.drawable.sample2, R.drawable.sample3,
        R.drawable.sample4, R.drawable.sample5,
        R.drawable.sample6, R.drawable.sample7,
        R.drawable.sample0, R.drawable.sample1,
        R.drawable.sample2, R.drawable.sample3,
        R.drawable.sample4, R.drawable.sample5,
        R.drawable.sample6, R.drawable.sample7
};
}
```

注意：当 Android 5.0 推出以后，提供了很多新东西，support v7 也更新了，出现了 AppCompatActivity。AppCompatActivity 是用来替代 ActionBarActivity 的，在本书中所有 AppCompatActivity 都可以手动改成 Activity，不会影响程序运行。

修改工程的 java/com.example.griddisplay 目录下 java 文件 MainActivity.java，修改代码如下：

```
package com.example.griddisplay;

import android.os.Bundle;
import android.support.v7.app.AppCompatActivity;
import android.widget.GridView;

public class MainActivity extends AppCompatActivity {

    @Override
    protected void onCreate(Bundle savedInstanceState) {
        super.onCreate(savedInstanceState);
        setContentView(R.layout.activity_main);
        GridView gridview = (GridView) findViewById(R.id.gridview);
        gridview.setAdapter(new ImageAdapter(this));
```

            }
    }

在 Genymotion 中运行程序的效果如图 2.12 所示。

图 2.12　GridView（网格视图）实例运行效果

【例 2.7】　GridView（网格视图）实例扩展

下面将【例 2.6】进行扩展：当单击网格视图中任意一个小图片时，打开一个新页面满屏显示这张图片。为了实现这一效果，需要新建一个 acitvity（注意：所有的 activity 都需要在 manifests/AndroidManifest.xml 文件中进行注册）。下面将通过表 2.9 中的步骤修改【例 2.6】来实现。

表 2.9　GridView 扩展实例步骤

步骤	描述
1	打开 Android Studio 选择 Open an existing Android Studio project（打开现有 Android Studio 工程），打开【例 2.6】的现有实例
2	在 java/com.example.griddisplay 目录下新建一个 Activity 类命名为 SingleViewActivity.java，修改代码如下所示
3	在 res/layout 目录下新建一个 XML 布局文件命名为 single_view.xml
4	在 manifests 目录下的 AndroidManifest.xml 文件中为新建的 activity 注册，添加代码如下所示
5	启动 Genymotion 模拟器，然后在 Android 工程中做如下代码修改

修改工程 res/layout 目录下新建的布局文件 single_view.xml，修改代码如下：

```
<?xml version="1.0" encoding="utf-8"?>
<LinearLayout xmlns:android="http://schemas.android.com/apk/res/android"
    android:layout_width="match_parent"
```

```xml
        android:layout_height="match_parent"
        android:orientation="vertical" >

    <ImageView android:id="@+id/SingleView"
        android:layout_width="fill_parent"
        android:layout_height="fill_parent"/>

</LinearLayout>
```

修改工程的 java/com.example.griddisplay 目录下新建的 Activity 类文件 SingleViewActivity.java，修改代码如下：

```java
package com.example.griddisplay;

import android.app.Activity;
import android.content.Intent;
import android.os.Bundle;
import android.widget.ImageView;

public class SingleViewActivity extends Activity {
    @Override
    public void onCreate(Bundle savedInstanceState) {
        super.onCreate(savedInstanceState);
        setContentView(R.layout.single_view);

        //Get intent data
        Intent i = getIntent();

        //Selected image id
        int position = i.getExtras().getInt("id");
        ImageAdapter imageAdapter = new ImageAdapter(this);

        ImageView imageView = (ImageView) findViewById(R.id.SingleView);
        imageView.setImageResource(imageAdapter.mThumbIds[position]);
    }
}
```

修改工程的 java/com.example.griddisplay 目录下 java 文件 MainActivity.java，修改代码如下：

```java
package com.example.griddisplay;

import android.content.Intent;
import android.os.Bundle;
import android.support.v7.app.AppCompatActivity;
import android.view.View;
import android.widget.AdapterView;
import android.widget.AdapterView.OnItemClickListener;
```

```java
import android.widget.GridView;

public class MainActivity extends AppCompatActivity {

    @Override
    protected void onCreate(Bundle savedInstanceState) {
        super.onCreate(savedInstanceState);
        setContentView(R.layout.activity_main);
        GridView gridview = (GridView) findViewById(R.id.gridview);
        gridview.setAdapter(new ImageAdapter(this));

        gridview.setOnItemClickListener(new OnItemClickListener() {
            public void onItemClick(AdapterView<?> parent, View v, int
                position, long id) {
                //Send intent to SingleViewActivity
                Intent i = new Intent(getApplicationContext(), SingleViewActivity.
                    class);

                //Pass image index
                i.putExtra("id", position);
                startActivity(i);
            }
        });
    }
}
```

修改工程的 manifests 目录下工程配置文件 AndroidManifest.xml，修改代码如下：

```xml
<?xml version="1.0" encoding="utf-8"?>
<manifest xmlns:android="http://schemas.android.com/apk/res/android"
    package="com.example.griddisplay" >

    <application
        android:allowBackup="true"
        android:icon="@mipmap/ic_launcher"
        android:label="@string/app_name"
        android:supportsRtl="true"
        android:theme="@style/AppTheme" >
        <activity android:name=".MainActivity" >
            <intent-filter>
                <action android:name="android.intent.action.MAIN" />

                <category android:name="android.intent.category.LAUNCHER" />
            </intent-filter>
        </activity>
        <activity android:name=".SingleViewActivity"></activity>
    </application>

</manifest>
```

在 Genymotion 中运行程序的效果如图 2.13 和图 2.14 所示。

图 2.13 GridView（网格视图）实例扩展运行效果 1

图 2.14 GridView（网格视图）实例扩展运行效果 2

2.4 Android UI 控件

Android 应用程序的人机交互界面由很多 Android 组件组成。Android 提供了很多 UI 控件，如图 2.15 中的部分常用 UI 控件：TextView（文本框）、Buttons（按钮）、CheckBox（多选框）、RadioButton（单选框）、ToggleButton（开关按钮）、RatingBar（星级评分条）等。

图 2.15 部分常用 UI 控件

2.4.1 TextView（文本框）

在 Android 中，TextView（文本框）用于在屏幕中显示文本。TextView 是一个简单 View，因此它是直接继承 View 对象。

- TextView 常用属性

TextView 的常用属性如表 2.10 所示，关于该控件的其他属性，可以参阅 Android 官方提供的完整的 API 文档。

表 2.10 TextView 的常用属性

属性	描述
android:id	用于设置文本框的名称
android:capitalize	用于设置输入英文字母是否自动变为大写字母 ● 0 表示不自动变为大写字母 ● 1 表示一句话的首字母自动变为大写字母 ● 2 表示每个单词的首字母自动变为大写字母 ● 3 表示所有字母自动变为大写字母
android:cursorVisible	当可以编辑时,光标是否显示 ● true 表示光标显示 ● false 表示光标不显示(默认值)
android:editable	如果设置为 true,表示文本框可以进行编辑
android:fontFamily	用于设置文本框的字体样式
android:gravity	用于设置文本框的对齐方式,可选值有 top、left、right、center_vertical、fill_vertical、center_horizontal、fill_horizontal、center、fill、clip_vertical 和 clip_horizontal 等。这些属性值也可以同时设定,各属性值之间用"|"(竖线)隔开。例如要指定组件靠右下角对齐,可以使用属性值 right|bottom
android:hint	用于设置当文本框中文本内容为空时,默认显示的提示文本
android:inputType	用于指定当前文本框显示内容的文本类型,其可选值有 textpassword、textEmailAddress、phone 和 date 等,可以同时指定多个,使用"|"(竖线)隔开。
android:maxHeight	用于设置文本框的最大高度
android:maxWidth	用于设置文本框的最大宽度
android:minHeight	用于设置文本框的最小高度
android:minWidth	用于设置文本框的最小宽度
android:password	用于设置当文本框可以编辑时,里面的文本是否显示为隐藏密码样式".",显示为 true,不显示为 false
android:phoneNumber	用于设置当文本框可以编辑时,里面的文本是否显示为电话号码样式,显示为 true,不显示为 false
android:text	用于指定文本框中显示的文本内容
android:textAllCaps	用于设置文本框中的英文字母是否全部变为大写字母,变成大写字母为 true,不变成大写字母为 false
android:textColor	用于设置文本框内文本的颜色,其属性值可以是"#rgb"、"#argb"、"#rrggbb"或 "#aarrggbb"
android:textColorHighlight	用于设置文本框内文本的颜色是否高亮
android:textColorHint	用于设置文本框提示信息的颜色,其属性值可以是"#rgb"、"#argb"、"#rrggbb"或 "#aarrggbb"
android:textIsSelectable	用于设置当文本框不可以编辑时,文本框中的文字是否可以被选中,true 为可以,false 为不可以
android:textSize	用于设置文本框中文本的字体大小,其属性由代表大小的数值和单位组成,其单位可以是 px、pt、sp 和 in 等
android:textStyle	用于设置文本框中文本的样式,可以同时指定多个,使用"|"(竖线)隔开 ● 0 表示正常 ● 1 表示加粗 ● 2 表示斜体

【例 2.8】 TextView(文本框)实例

这个例子将使用 TextView(文本框)控件来创建一个简单的 Android 应用,下面将通过表 2.11 中的步骤来实现。

表 2.11 TextView(文本框)实例步骤

步骤	描述
1	打开 Android Studio 创建一个 Android 应用,Application name(应用名)取名为 TextView,Company Domain(公司域)取名为 Example.com
2	Minimum SDK 选择 API 18: Android 4.3(Jelly Bean)
3	选择 Empty Activity
4	不需要修改 Activity Name,使用默认值,单击"Finish"按钮(完成)
5	在工程中找到 ras/layout 目录中的 activity_main.xml 文件,在其中添加一个 TextView 控件
6	启动 Genymotion 模拟器,然后在 Android 工程中做如下代码修改

修改新建工程 res/layout 目录下的布局文件 activity_main.xml，修改代码如下：

```xml
<?xml version="1.0" encoding="utf-8"?>
<RelativeLayout xmlns:android="http://schemas.android.com/apk/res/android"
    xmlns:tools="http://schemas.android.com/tools" android:layout_width=
               "match_parent"
    android:layout_height="match_parent" android:paddingLeft="@dimen/activity_
                        horizontal_margin"
    android:paddingRight="@dimen/activity_horizontal_margin"
    android:paddingTop="@dimen/activity_vertical_margin"
    android:paddingBottom="@dimen/activity_vertical_margin" tools:context=
                        ".MainActivity">

    <TextView
        android:id="@+id/text_id"
        android:layout_width="300dp"
        android:layout_height="200dp"
        android:capitalize="characters"
        android:text="hello_world"
        android:textColor="@android:color/holo_blue_dark"
        android:textColorHighlight="@android:color/primary_text_dark"
        android:layout_centerVertical="true"
        android:layout_alignParentEnd="true"
        android:textSize="50dp"/>
</RelativeLayout>
```

在 Genymotion 中运行程序的效果如图 2.16 所示。

图 2.16　TextView（文本框）实例运行效果

2.4.2 EditText（编辑框）

EditText（编辑框）用于在屏幕中显示文本输入框。在 Android 中 EditText（编辑框）控件可以输入单行文本，也可以输入多行文本，还可以输入指定格式的文本（如密码、电话号码、E-mail 地址等）。

● EditText 的常用属性

下面介绍 EditText（编辑框）的常用属性，关于该控件的其他属性，可以参阅 Android 官方提供的完整的 API 文档。

(1) 继承于 android.widget.TextView 类的 EditText 的常用属性如表 2.12 所示。

表 2.12　继承于 android.widget.TextView 类的 EditText 的常用属性

属性	描述
android:autoText	被选中表示对编辑框中的文本自动更正拼写错误
android:drawableBottom	用于在编辑框内文本的底端绘制指定图片，该图片可以是放在 res/drawable 目录下的图片，通过"@drawable/文件名(不包括文件的扩展名)"设置
android:drawableRight	用于在编辑框内文本的右侧绘制指定图片，该图片可以是放在 res/drawable 目录下的图片，通过"@drawable/文件名(不包括文件的扩展名)"设置
android:editable	用于设置编辑框是否能被编辑，默认值为 true
android:text	用于设置编辑框中默认显示的文本

(2) 继承于 android.view.View 类的 EditText 的常用属性如表 2.13 所示。

表 2.13　继承于 android.view.View 类的 EditText 的常用属性

属性	描述
android:background	用于设置编辑框的背景图片，该图片可以是放在 res/drawable 目录下的图片，通过"@drawable/文件名(不包括文件的扩展名)"设置
android:contentDescription	用于设置编辑框的简单描述文字
android:id	用于设置编辑框的名称
android:onClick	用于设置单击事件响应方法的方法名
android:visibility	用于设置编辑框是否可见

【例 2.9】　EditText（编辑框）实例

这个例子将使用 EditText（编辑框）控件来创建一个简单的 Android 应用，下面将通过表 2.14 中的步骤来实现。

表 2.14　EditText（编辑框）实例步骤

步骤	描述
1	打开 Android Studio 创建一个 Android 应用，Application name（应用名）取名为 EditText，Company Domain（公司域）取名为 Example.com
2	Minimum SDK 选择 API 18: Android 4.3 (Jelly Bean)
3	选择 Empty Activity
4	不需要修改 Activity Name，使用默认值，单击"Finish"按钮（完成）
5	在工程中找到 ras/layout 目录中的 activity_main.xml 文件，在其中添加一个 EditText（编辑框）控件
6	启动 Genymotion 模拟器，然后在 Android 工程中做如下代码修改

修改新建工程 res/layout 目录下的布局文件 activity_main.xml，修改代码如下：

```xml
<?xml version="1.0" encoding="utf-8"?>
<RelativeLayout xmlns:android="http://schemas.android.com/apk/res/android"
    xmlns:tools="http://schemas.android.com/tools" android:layout_width=
                "match_parent"
    android:layout_height="match_parent" android:paddingLeft="@dimen/
                        activity_horizontal_margin"
    android:paddingRight="@dimen/activity_horizontal_margin"
    android:paddingTop="@dimen/activity_vertical_margin"
    android:paddingBottom="@dimen/activity_vertical_margin" tools:context=
                ".MainActivity">

    <TextView
        android:id="@+id/textView1"
        android:layout_width="wrap_content"
        android:layout_height="wrap_content"
        android:layout_alignParentLeft="true"
        android:layout_alignParentTop="true"
        android:layout_marginLeft="14dp"
        android:layout_marginTop="18dp"
        android:text="@string/example_edittext" />

    <Button
        android:id="@+id/button"
        android:layout_width="wrap_content"
        android:layout_height="wrap_content"
        android:layout_alignLeft="@+id/textView1"
        android:layout_below="@+id/textView1"
        android:layout_marginTop="130dp"
        android:text="@string/show_the_text" />

    <EditText
        android:id="@+id/edittext"
        android:layout_width="fill_parent"
        android:layout_height="wrap_content"
        android:layout_alignLeft="@+id/button"
        android:layout_below="@+id/textView1"
        android:layout_marginTop="61dp"
        android:ems="10"
        android:text="@string/enter_text" android:inputType="text" />
</RelativeLayout>
```

修改新建工程 values 目录下的字符串文件 strings.xml，修改代码如下：

```xml
<?xml version="1.0" encoding="utf-8"?>
<resources>
    <string name="app_name">demo</string>
    <string name="action_settings">Settings</string>
    <string name="example_edittext">Example showing EditText</string>
```

```
    <string name="show_the_text">Show the Text</string>
    <string name="enter_text">text changes</string>
</resources>
```

在 Genymotion 中运行程序的效果如图 2.17 所示。

图 2.17 EditText(编辑框)实例运行效果

2.4.3 AutoCompleteTextView(自动填充文本框)

AutoCompleteTextView(自动填充文本框)用于实现允许用户输入一定的字符后,显示一个下拉菜单,供用户从中选择,当用户选择某个选项后,按用户选择自动填写文本框。

● AutoCompleteTextView 的常用属性

AutoCompleteTextView 的常用属性如表 2.15 所示,关于该控件的其他属性,可以参阅 Android 官方提供的完整的 API 文档。

表 2.15 AutoCompleteTextView 的常用属性

属性	描述
android:completionHint	用于为弹出的下拉菜单指定提示标题
android:completionHintView	用于设置弹出的下拉菜单底部信息的样式
android:completionThreshold	用于指定用户至少输入几个字符才会显示提示
android:dropDownAnchor	它的值是一个 View 的 ID,指定后,AutoCompleteTextView 会在这个 View 下弹出自动提示
android:dropDownHeight	用于指定下拉菜单的高度
android:dropDownHorizontalOffset	用于指定下拉菜单与文本之间的水平偏移。下拉菜单默认与文本框左对齐
android:dropDownSelector	用于设置下拉菜单的显示效果
android:dropDownVerticalOffset	用于设置下拉菜单与文本之间的垂直偏移。下拉菜单默认紧跟文本框
android:dropDownWidth	用于设置下拉菜单的宽度
android:popupBackground	用于设置下拉菜单的背景

【例 2.10】 AutoCompleteTextView（自动填充文本框）实例

这个例子将使用 AutoCompleteTextView（自动填充文本框）控件来创建一个简单的 Android 应用，下面将通过表 2.16 中的步骤来实现。

表 2.16　AutoCompleteTextView（自动填充文本框）实例步骤

步骤	描述
1	打开 Android Studio 创建一个 Android 应用，Application name（应用名）取名为 AutoCompleteTextView，Company Domain（公司域）取名为 Example.com
2	Minimum SDK 选择 API 18: Android 4.3 (Jelly Bean)
3	选择 Empty Activity
4	不需要修改 Activity Name，使用默认值，单击 Finish（完成）
5	在工程中找到 ras/layout 目录中的 activity_main.xml 文件，在其中添加一个 AutoCompleteTextView（自动填充文本框）控件
6	启动 Genymotion 模拟器，然后在 Android 工程中做如下代码修改

修改新建工程 res/layout 目录下的布局文件 activity_main.xml，修改代码如下：

```xml
<?xml version="1.0" encoding="utf-8"?>
<RelativeLayout xmlns:android="http://schemas.android.com/apk/res/android"
    xmlns:tools="http://schemas.android.com/tools" android:layout_width=
        "match_parent"
    android:layout_height="match_parent" android:paddingLeft="@dimen/
        activity_horizontal_margin"
    android:paddingRight="@dimen/activity_horizontal_margin"
    android:paddingTop="@dimen/activity_vertical_margin"
    android:paddingBottom="@dimen/activity_vertical_margin" tools:context=
        ".MainActivity">

    <TextView
        android:id="@+id/textView2"
        android:layout_width="wrap_content"
        android:layout_height="wrap_content"
        android:layout_alignParentTop="true"
        android:layout_centerHorizontal="true"
        android:layout_marginTop="25dp"
        android:text="Example showing AutoCompleteTextView" />

    <AutoCompleteTextView
        android:id="@+id/autoCompleteTextView1"
        android:layout_width="wrap_content"
        android:layout_height="wrap_content"
        android:layout_alignLeft="@+id/textView2"
        android:layout_below="@+id/textView2"
        android:layout_marginTop="54dp"
        android:ems="10" />
</RelativeLayout>
```

修改新建工程 java/com.example.autocompletetextview 目录下的 java 文件 MainActivity.java，修改代码如下：

```java
package com.example.autocompletetextview;

import android.os.Bundle;
import android.support.v7.app.AppCompatActivity;
import android.widget.ArrayAdapter;
import android.widget.AutoCompleteTextView;

public class MainActivity extends AppCompatActivity {

    AutoCompleteTextView autocomplete;

    String[] arr = { "Paries,France", "PA,United States","Parana,Brazil",
                    "Padua,Italy", "Pasadena,CA,United States"};

    @Override
    protected void onCreate(Bundle savedInstanceState) {
        super.onCreate(savedInstanceState);
        setContentView(R.layout.activity_main);

        autocomplete = (AutoCompleteTextView)
                findViewById(R.id.autoCompleteTextView1);

        ArrayAdapter<String> adapter = new ArrayAdapter<String>
                (this,android.R.layout.select_dialog_item, arr);

        autocomplete.setThreshold(2);
        autocomplete.setAdapter(adapter);
    }
}
```

在 AutoCompleteTextView(自动填充文本框)控件中输入"pa",自动提示下拉菜单在 Genymotion 中运行效果如图 2.18 所示。

图 2.18 AutoCompleteTextView(自动填充文本框)实例运行效果

2.4.4 Button(普通按钮)

Button(普通按钮)用于在 UI 界面上生成一个可以单击的按钮。当用户单击按钮时，将会触发一个 onClick 事件，可以通过为按钮添加的单击事件监听器指定所要触发的动作。

● Button 的常用属性

下面介绍 Button(普通按钮)的常用属性，关于该控件的其他属性，可以参阅 Android 官方提供的完整的 API 文档。

(1) 继承于 android.widget.TextView 类的 Button 的常用属性如表 2.17 所示。

表 2.17 继承于 android.widget.TextView 类的 Button 的常用属性

属性	描述
android:autoText	被选中表示对按钮上显示的文本进行自动更正拼写错误
android:drawableBottom	用于在按钮上显示文本的底端绘制指定图片，该图片可以是放在 res/drawable 目录下的图片，通过"@drawable/文件名(不包括文件的扩展名)"设置
android:drawableRight	用于在按钮上显示文本的右侧绘制指定图片，该图片可以是放在 res/drawable 目录下的图片，通过"@drawable/文件名(不包括文件的扩展名)"设置
android:enabled	用于设置按钮是否可用
android:text	用于设置按钮上显示的文字

(2) 继承于 android.view.View 类的 Button 的常用属性如表 2.18 所示。

表 2.18 继承于 android.view.View 类的 Button 的常用属性

属性	描述
android:background	用于设置按钮的背景图片，该图片可以是放在 res/drawable 目录下的图片，通过"@drawable/文件名(不包括文件的扩展名)"设置
android:contentDescription	用于设置按钮的简单描述文字
android:id	用于设置按钮的名称
android:onClick	用于设置单击事件响应方法的方法名
android:visibility	用于设置按钮控件是否可见

【例 2.11】 Button(普通按钮)实例

这个例子将使用 Button(普通按钮)控件来创建一个简单的 Android 应用，下面将通过表 2.19 中的步骤来实现。

表 2.19 Button(普通按钮)实例步骤

步骤	描述
1	打开 Android Studio 创建一个 Android 应用，Application name(应用名)取名为 Button，Company Domain(公司域)取名为 Example.com
2	Minimum SDK 选择 API 18: Android 4.3 (Jelly Bean)
3	选择 Empty Activity
4	不需要修改 Activity Name，使用默认值，单击"Finish"按钮(完成)
5	在工程中找到 ras/layout 目录中的 activity_main.xml 文件，在其中添加一个 TextView(文本框)控件、一个 ImageButton(图片按钮)、一个 EditText(编辑框)和一个 Button(普通按钮)控件，按图 2.19 所示的运行效果位置进行摆放
6	将工程 res/mipmap 目录下的 ic_launcher.png 图片文件复制、粘贴到 res/drawable 目录下
7	启动 Genymotion 模拟器，然后在 Android 工程中做如下代码修改

修改新建工程 res/layout 目录下的布局文件 activity_main.xml，修改代码如下：

```xml
<?xml version="1.0" encoding="utf-8"?>
<RelativeLayout xmlns:android="http://schemas.android.com/apk/res/android"
    xmlns:tools="http://schemas.android.com/tools" android:layout_width=
              "match_parent"
    android:layout_height="match_parent" android:paddingLeft="@dimen/
                          activity_horizontal_margin"
    android:paddingRight="@dimen/activity_horizontal_margin"
    android:paddingTop="@dimen/activity_vertical_margin"
    android:paddingBottom="@dimen/activity_vertical_margin" tools:context=
                          ".MainActivity">

    <TextView
        android:id="@+id/textView1"
        android:layout_width="wrap_content"
        android:layout_height="wrap_content"
        android:text="Button Control"
        android:layout_alignParentTop="true"
        android:layout_centerHorizontal="true"
        android:textSize="30dp" />

    <ImageButton
        android:layout_width="wrap_content"
        android:layout_height="wrap_content"
        android:id="@+id/imageButton"
        android:src="@drawable/ic_launcher"
        android:layout_centerHorizontal="true"
        android:layout_below="@+id/textView1" />

    <Button
        android:layout_width="wrap_content"
        android:layout_height="wrap_content"
        android:text="Button"
        android:id="@+id/button"
        android:layout_alignStart="@+id/imageButton"
        android:layout_below="@+id/imageButton"
        android:layout_alignRight="@+id/imageButton"
        android:layout_alignEnd="@+id/imageButton" />
</RelativeLayout>
```

修改新建工程 java/com.example.button 目录下的 java 文件 MainActivity.java，修改代码如下：

```java
package com.example.button;

import android.app.Activity;
import android.os.Bundle;
import android.view.View;
import android.widget.Button;
```

```
    import android.widget.Toast;

public class MainActivity extends Activity {
    Button b1;
    @Override
    protected void onCreate(Bundle savedInstanceState) {
        super.onCreate(savedInstanceState);
        setContentView(R.layout.activity_main);

        b1=(Button)findViewById(R.id.button);
        b1.setOnClickListener(new View.OnClickListener() {
            @Override
            public void onClick(View v) {
                Toast.makeText(MainActivity.this, "YOUR MESSAGE", Toast.
                        LENGTH_LONG).show();
            }
        });
    }
}
```

在 Genymotion 中运行，然后单击 Button（普通按钮），其效果如图 2.19 所示。

图 2.19　Button（普通按钮）实例运行效果图

2.4.5　ImageButton（图片按钮）

图片按钮和普通按钮的使用方法基本相同，只不过图片按钮有 android:src 属性，可以用来设置在按钮中显示的图片。与普通按钮一样，也需要为图片按钮添加单击事件监听器用于指定所要触发的动作。

- ImageButton 的常用属性

下面介绍 ImageButton 的常用属性，关于该控件的其他属性，可以参阅 Android 官方提供的完整的 API 文档。

(1) 继承于 android.widget.ImageView 类的 ImageButton 的常用属性如表 2.20 所示。

表 2.20　继承于 android.widget.ImageView 类的 ImageButton 的常用属性

属性	描述
android:adjustViewBounds	如果设置为 true，ImageButton 将调整其边框以适应图片大小
android:baseline	用于设置 ImageButton 的基线
android:baselineAlignBottom	值为 true 表示图片的基线与按钮底部相对齐
android:cropToPadding	值为 true 表示 ImageButton 会剪切图片以适应内边距的大小
android:src	用于设置要显示的图片资源位置，该图片可以是放在 res/drawable 目录下的图片，通过"@drawable/文件名(不包括文件的扩展名)"设置

(2) 继承于 android.view.View 类的 ImageButton 的常用属性如表 2.21 所示。

表 2.21　继承于 android.view.View 类的 ImageButton 的常用属性

属性	描述
android:background	用于设置图片按钮的背景图片，该图片可以是放在 res/drawable 目录下的图片，通过"@drawable/文件名(不包括文件的扩展名)"设置
android:contentDescription	用于设置图片按钮的简单描述文字
android:id	用于设置按钮的名称
android:onClick	用于设置单击事件响应方法的方法名
android:visibility	用于设置按钮控件是否可见

【例 2.12】　ImageButton(图片按钮)实例

这个例子将使用 ImageButton(图片按钮)控件来创建一个简单的 Android 应用，下面将通过表 2.22 中的步骤来实现。

表 2.22　ImageButton(图片按钮)实例步骤

步骤	描述
1	打开 Android Studio 创建一个 Android 应用，Application name(应用名)取名为 ImageButton，Company Domain(公司域)取名为 Example.com
2	Minimum SDK 选择 API 18: Android 4.3 (Jelly Bean)
3	选择 Empty Activity
4	不需要修改 Activity Name，使用默认值，单击"Finish"按钮(完成)
5	在工程中找到 ras/layout 目录中的 activity_main.xml 文件，在其中添加一个 TextView(文本框)控件、一个 ImageButton(图片按钮)，按如图 2.20 所示的运行效果位置进行摆放
6	将工程 res/mipmap 目录下的 ic_launcher.png 图片文件复制、粘贴到 res/drawable 目录下
7	启动 Genymotion 模拟器，然后在 Android 工程中做如下代码修改

修改新建工程 res/layout 目录下的布局文件 activity_main.xml，修改代码如下：

```
<?xml version="1.0" encoding="utf-8"?>
<RelativeLayout xmlns:android="http://schemas.android.com/apk/res/android"
    xmlns:tools="http://schemas.android.com/tools" android:layout_width=
        "match_parent"
    android:layout_height="match_parent" android:paddingLeft="@dimen/activity_
```

```xml
                         horizontal_margin"
    android:paddingRight="@dimen/activity_horizontal_margin"
    android:paddingTop="@dimen/activity_vertical_margin"
    android:paddingBottom="@dimen/activity_vertical_margin" tools:context=
                ".MainActivity">

    <TextView android:text="ImageButton"
        android:layout_width="wrap_content"
        android:layout_height="wrap_content"
        android:textSize="30dp"
        android:layout_alignParentTop="true"
        android:layout_alignRight="@+id/imageButton"
        android:layout_alignEnd="@+id/imageButton" />

    <ImageButton
        android:layout_width="wrap_content"
        android:layout_height="wrap_content"
        android:id="@+id/imageButton"
        android:layout_centerVertical="true"
        android:layout_centerHorizontal="true"
        android:src="@drawable/ic_launcher"/>
</RelativeLayout>
```

修改新建工程 java/com.example.imagebutton 目录下的 java 文件 MainActivity.java，修改代码如下：

```java
package com.example.imagebutton;

import android.os.Bundle;
import android.support.v7.app.AppCompatActivity;
import android.view.View;
import android.widget.ImageButton;
import android.widget.Toast;

public class MainActivity extends AppCompatActivity {
    ImageButton imgButton;
    @Override
    protected void onCreate(Bundle savedInstanceState) {
        super.onCreate(savedInstanceState);
        setContentView(R.layout.activity_main);
        imgButton =(ImageButton)findViewById(R.id.imageButton);
        imgButton.setOnClickListener(new View.OnClickListener() {
            @Override
            public void onClick(View v) {
                Toast.makeText(getApplicationContext(), "You download is
                        resumed", Toast.LENGTH_LONG).show();
            }
```

```
        });
    }
}
```

在 Genymotion 中运行，然后单击 ImageButton(图片按钮)，其效果如图 2.20 所示。

图 2.20　ImageButton(图片按钮)实例运行效果

2.4.6　CheckBox(复选框)

在 Android 中，单选按钮和复选框都继承于普通按钮，因此，它们都可以直接使用普通按钮支持的各种属性和方法，与普通按钮不同的是，它们提供了可选中的功能。

在默认情况下，CheckBox(复选框)显示为一个方框图标，并且在该图标旁边放置一些说明文字。CheckBox 可以进行多选设置，每个复选框都提供"选中"和"不选中"两种状态。

● CheckBox 的常用属性

下面介绍 CheckBox 的常用属性，关于该控件的其他属性，可以参阅 Android 官方提供的完整的 API 文档。

(1)继承于 android.widget.TextView 类的 CheckBox(复选框)的常用属性如表 2.23 所示。

表 2.23　继承于 android.widget.TextView 类的 CheckBox(复选框)的常用属性

属性	描述
android:autoText	被选中表示对复选框的说明文本自动更正拼写错误
android:drawableBottom	用于在复选框的说明文本的底端绘制指定图片，该图片可以是放在 res/drawable 目录下的图片，通过"@drawable/文件名(不包括文件的扩展名)"设置
android:drawableRight	用于在复选框的说明文本的右侧绘制指定图片，该图片可以是放在 res/drawable 目录下的图片，通过"@drawable/文件名(不包括文件的扩展名)"设置
android:editable	用于设置复选框的说明文本是否能被编辑，默认值为 true
android:text	用于设置复选框的默认说明文本

(2)继承于 android.view.View 类的 CheckBox(复选框)的常用属性如表 2.24 所示。

表 2.24 继承于 android.view.View 类的 CheckBox(复选框)的常用属性

属性	描述
android:background	用于设置复选框的背景图片,该图片可以是放在 res/drawable 目录下的图片,通过"@drawable/文件名(不包括文件的扩展名)"设置
android:contentDescription	用于设置复选框的简单描述文字
android:id	用于设置复选框的名称
android:onClick	用于设置单击事件响应方法的方法名
android:visibility	用于设置复选框是否可见

【例 2.13】 CheckBox(复选框)实例

这个例子将使用 CheckBox(复选框)控件来创建一个简单的 Android 应用,下面将通过表 2.25 中的步骤来实现。

表 2.25 CheckBox(复选框)实例步骤

步骤	描述
1	打开 Android Studio 创建一个 Android 应用,Application name(应用名)取名为 CheckBox,Company Domain(公司域)取名为 Example.com
2	Minimum SDK 选择 API 18: Android 4.3 (Jelly Bean)
3	选择 Empty Activity
4	不需要修改 Activity Name,使用默认值,单击"Finish"按钮(完成)
5	在工程中找到 ras/layout 目录中的 activity_main.xml 文件,在其中添加一个 TextView(文本框)控件、两个 CheckBox(复选框)、两个 Button(普通按钮)和一个 ImageButton(图片按钮),布局按图 2.21 所示的运行效果位置进行摆放
6	将工程 res/mipmap 目录下 ic_launcher.png 图片文件复制、粘贴到 res/drawable 目录下
7	启动 Genymotion 模拟器,然后在 Android 工程中做如下代码修改

修改新建工程 res/layout 目录下的布局文件 activity_main.xml,修改代码如下:

```xml
<?xml version="1.0" encoding="utf-8"?>
<RelativeLayout xmlns:android="http://schemas.android.com/apk/res/android"
    xmlns:tools="http://schemas.android.com/tools" android:layout_width=
        "match_parent"
    android:layout_height="match_parent" android:paddingLeft="@dimen/
            activity_horizontal_margin"
    android:paddingRight="@dimen/activity_horizontal_margin"
    android:paddingTop="@dimen/activity_vertical_margin"
    android:paddingBottom="@dimen/activity_vertical_margin" tools:context=
        ".MainActivity">
    <TextView
        android:id="@+id/textView1"
        android:layout_width="wrap_content"
        android:layout_height="wrap_content"
        android:text="Example of checkbox"
        android:layout_alignParentTop="true"
        android:layout_centerHorizontal="true"
        android:textSize="30dp" />
```

```xml
<CheckBox
    android:id="@+id/checkBox1"
    android:layout_width="wrap_content"
    android:layout_height="wrap_content"
    android:text="Do you like Tutorials Point"
    android:layout_above="@+id/button"
    android:layout_centerHorizontal="true" />

<CheckBox
    android:id="@+id/checkBox2"
    android:layout_width="wrap_content"
    android:layout_height="wrap_content"
    android:text="Do you like android "
    android:checked="false"
    android:layout_above="@+id/checkBox1"
    android:layout_alignLeft="@+id/checkBox1"
    android:layout_alignStart="@+id/checkBox1" />

<Button
    android:layout_width="wrap_content"
    android:layout_height="wrap_content"
    android:text="Ok"
    android:id="@+id/button"
    android:layout_alignParentBottom="true"
    android:layout_alignLeft="@+id/checkBox1"
    android:layout_alignStart="@+id/checkBox1" />

<Button
    android:layout_width="wrap_content"
    android:layout_height="wrap_content"
    android:text="Cancel"
    android:id="@+id/button2"
    android:layout_alignParentBottom="true"
    android:layout_alignRight="@+id/textView1"
    android:layout_alignEnd="@+id/textView1" />

<ImageButton
    android:layout_width="wrap_content"
    android:layout_height="wrap_content"
    android:id="@+id/imageButton"
    android:src="@drawable/ic_launcher"
    android:layout_centerVertical="true"
    android:layout_centerHorizontal="true" />
</RelativeLayout>
```

修改新建工程 java/com.example.checkbox 目录下的 java 文件 MainActivity.java，修改代码如下：

```java
package com.example.checkbox;

import android.os.Bundle;
import android.support.v7.app.AppCompatActivity;
import android.view.View;
import android.widget.Button;
import android.widget.CheckBox;
import android.widget.Toast;

public class MainActivity extends AppCompatActivity {
    CheckBox ch1,ch2;
    Button b1,b2;
    @Override
    protected void onCreate(Bundle savedInstanceState) {
        super.onCreate(savedInstanceState);
        setContentView(R.layout.activity_main);
        ch1=(CheckBox)findViewById(R.id.checkBox1);
        ch2=(CheckBox)findViewById(R.id.checkBox2);

        b1=(Button)findViewById(R.id.button);
        b2=(Button)findViewById(R.id.button2);
        b2.setOnClickListener(new View.OnClickListener() {
            @Override
            public void onClick(View v) {
                finish();
            }
        });

        b1.setOnClickListener(new View.OnClickListener() {
            @Override
            public void onClick(View v) {
                StringBuffer result = new StringBuffer();
                result.append("Thanks : ").append(ch1.isChecked());
                result.append("\nThanks: ").append(ch2.isChecked());
                Toast.makeText(MainActivity.this, result.toString(), Toast.
                        LENGTH_LONG).show();
            }
        });
    }
}
```

在 Genymotion 中运行，然后选中其中一个 CheckBox（复选框）或同时选中两个 CheckBox（复选框），并单击"OK"按钮，效果如图 2.21 所示。

第 2 章　Android UI 设计

图 2.21　CheckBox（复选框）实例运行效果

2.4.7　ToggleButton（开关按钮）

ToggleButton（开关按钮）是 Android 中比较简单的一个控件，具有选中和未选中两个状态的按钮，并且需要为不同的状态设置不同的显示文本，同时还有一个显示开/关的指示灯。

● ToggleButton 的常用属性

ToggleButton 的常用属性如表 2.26 所示。关于该控件的其他属性，可以参阅 Android 官方提供的完整的 API 文档。

表 2.26　ToggleButton 的常用属性 1

属性	描述
android:disabledAlpha	用于设置按钮在禁用时的透明度
android:textOff	用于设置开关按钮未被选中时的说明文本
android:textOn	用于设置开关按钮被选中时的说明文本

继承于 android.widget.TextView 类的 ToggleButton 的常用属性如表 2.27 所示。

表 2.27　ToggleButton 的常用属性 2

属性	描述
android:autoText	被选中表示对开关按钮的说明文本自动更正拼写错误
android:drawableBottom	用于在开关按钮的说明文本的底端绘制指定图片，该图片可以是放在 res/drawable 目录下的图片，通过"@drawable/文件名(不包括文件的扩展名)"设置
android:drawableRight	用于在开关按钮的说明文本的右侧绘制指定图片，该图片可以是放在 res/drawable 目录下的图片，通过"@drawable/文件名(不包括文件的扩展名)"设置
android:editable	用于设置开关按钮的说明文本是否能被编辑，默认值为 true
android:text	用于设置开关按钮的默认说明文本

继承于 android.view.View 类的 ToggleButton 的常用属性如表 2.28 所示。

表 2.28　继承于 android.view.View 类的 ToggleButton 的常用属性

属性	描述
android:background	用于设置开关按钮的背景图片，该图片可以是放在 res/drawable 目录下的图片，通过"@drawable/文件名(不包括文件的扩展名)"设置
android:contentDescription	用于设置开关按钮的简单描述文字
android:id	用于设置开关按钮的名称
android:onClick	用于设置单击事件响应方法的方法名
android:visibility	用于设置开关按钮是否可见

【例 2.14】　ToggleButton(开关按钮)实例

这个例子将使用 ToggleButton(开关按钮)控件来创建一个简单的 Android 应用，下面将通过表 2.29 中的步骤来实现。

表 2.29　ToggleButton(开关按钮)实例步骤

步骤	描述
1	打开 Android Studio 创建一个 Android 应用，Application name(应用名)取名为 ToggleButton，Company Domain(公司域)取名为 Example.com
2	Minimum SDK 选择 API 18: Android 4.3 (Jelly Bean)
3	选择 Empty Activity
4	不需要修改 Activity Name，使用默认值，单击"Finish"按钮(完成)
5	在工程中找到 ras/layout 目录中的 activity_main.xml 文件，在其中添加一个 TextView(文本框)控件、两个 CheckBox(复选框)、两个 Button(普通按钮)和一个 ImageButton(图片按钮)，布局按图 2.22 所示的运行效果位置进行摆放
6	将工程 res/mipmap 目录下 ic_launcher.png 图片文件复制、粘贴到 res/drawable 目录下
7	启动 Genymotion 模拟器，然后在 Android 工程中做如下代码修改

修改新建工程 res/layout 目录下的布局文件 activity_main.xml，修改代码如下：

```
<?xml version="1.0" encoding="utf-8"?>
<RelativeLayout xmlns:android="http://schemas.android.com/apk/res/android"
    xmlns:tools="http://schemas.android.com/tools" android:layout_width=
        "match_parent"
    android:layout_height="match_parent" android:paddingLeft="@dimen/
        activity_horizontal_margin"
    android:paddingRight="@dimen/activity_horizontal_margin"
```

```xml
        android:paddingTop="@dimen/activity_vertical_margin"
        android:paddingBottom="@dimen/activity_vertical_margin" tools:context=
                ".MainActivity">

    <TextView
        android:id="@+id/textView1"
        android:layout_width="wrap_content"
        android:layout_height="wrap_content"
        android:text="Tutorials point"
        android:textColor="#ff87ff09"
        android:textSize="30dp"
        android:layout_above="@+id/imageButton"
        android:layout_centerHorizontal="true"
        android:layout_marginBottom="40dp" />

    <ImageButton
        android:layout_width="wrap_content"
        android:layout_height="wrap_content"
        android:id="@+id/imageButton"
        android:src="@drawable/ic_launcher"
        android:layout_centerVertical="true"
        android:layout_centerHorizontal="true" />

    <ToggleButton
        android:layout_width="wrap_content"
        android:layout_height="wrap_content"
        android:text="On"
        android:id="@+id/toggleButton"
        android:checked="true"
        android:layout_below="@+id/imageButton"
        android:layout_toLeftOf="@+id/imageButton"
        android:layout_toStartOf="@+id/imageButton" />

    <ToggleButton
        android:layout_width="wrap_content"
        android:layout_height="wrap_content"
        android:text="Off"
        android:id="@+id/toggleButton2"
        android:checked="true"
        android:layout_below="@+id/imageButton"
        android:layout_toEndOf="@+id/imageButton" />

    <Button
        android:layout_width="wrap_content"
        android:layout_height="wrap_content"
        android:id="@+id/button2"
```

```
            android:text="ClickMe"
            android:layout_alignParentBottom="true"
            android:layout_centerHorizontal="true" />
</RelativeLayout>
```

修改新建工程 java/com.example.togglebutton 目录下的 java 文件 MainActivity.java，修改代码如下：

```java
package com.example.togglebutton;

import android.os.Bundle;
import android.support.v7.app.AppCompatActivity;
import android.view.View;
import android.widget.Button;
import android.widget.Toast;
import android.widget.ToggleButton;

public class MainActivity extends AppCompatActivity {
    ToggleButton tg1,tg2;
    Button b1;
    @Override
    protected void onCreate(Bundle savedInstanceState) {
        super.onCreate(savedInstanceState);
        setContentView(R.layout.activity_main);
        tg1=(ToggleButton)findViewById(R.id.toggleButton);
        tg2=(ToggleButton)findViewById(R.id.toggleButton2);

        b1=(Button)findViewById(R.id.button2);
        b1.setOnClickListener(new View.OnClickListener() {
            @Override
            public void onClick(View v) {
                StringBuffer result = new StringBuffer();
                result.append("You have clicked first ON Button-:) ").
                        append(tg1.getText());
                result.append("You have clicked Second ON Button -:) ").
                        append(tg2.getText());
                Toast.makeText(MainActivity.this, result.toString(), Toast.
                        LENGTH_SHORT).show();
            }
        });
    }
}
```

在 Genymotion 中运行，然后单击其中一个 ToggleButton（开关按钮）使其状态变为 off，并单击"CLICKME"按钮，效果如图 2.22 所示。

图 2.22 ToggleButton（开关按钮）实例运行效果

2.4.8 RadioButton（单选按钮）与 RadioGroup（按钮组）

在默认情况下，单选按钮显示为一个圆形图标，并且在该图标旁边放置一些说明文字。一般将多个单选按钮放置在 RadioGroup（按钮组）中，当用户选中其中某个 RadioButton（单选按钮）后，RadioGroup 中的其他按钮将被自动取消选中状态，RadioButton 是 Button（普通按钮）类的子类，所以可以直接使用 Button 的各种属性。

● RadioButton 的常用属性

下面介绍 RadioButton 的常用属性，关于该控件的其他属性，可以参阅 Android 官方提供的完整的 API 文档。

（1）继承于 android.widget.TextView 类的 RadioButton 的常用属性如表 2.30 所示。

表 2.30 继承于 android.widget.TextView 类的 RadioButton 的常用属性

属性	描述
android:autoText	被选中表示对单选按钮的说明文本上显示的文本进行自动更正拼写错误
android:drawableBottom	用于在单选按钮的说明文本的底端绘制指定图片，该图片可以是放在 res/drawable 目录下的图片，通过"@drawable/文件名(不包括文件的扩展名)"设置
android:drawableRight	用于在单选按钮的说明文本的右侧绘制指定图片，该图片可以是放在 res/drawable 目录下的图片，通过"@drawable/文件名(不包括文件的扩展名)"设置
android:enabled	用于设置单选按钮是否可用
android:text	用于设置单选按钮的说明文本

（2）继承于 android.view.View 类的 RadioButton 的常用属性如表 2.31 所示。

表 2.31 继承于 android.view.View 类的 RadioButton 的常用属性

属性	描述
android:background	用于设置单选按钮的背景图片，该图片可以是放在 res/drawable 目录下的图片，通过"@drawable/文件名(不包括文件的扩展名)"设置
android:contentDescription	用于设置单选按钮的简单描述文字
android:id	用于设置单选按钮的名称
android:onClick	用于设置单击事件响应方法的方法名
android:visibility	用于设置单选按钮控件是否可见

【例 2.15】 RadioButton（单选按钮）实例

这个例子将使用 RadioButton（单选按钮）控件来创建一个简单的 Android 应用，下面将通过表 2.32 中的步骤来实现。

表 2.32　RadioButton（单选按钮）实例步骤

步骤	描述
1	打开 Android Studio 创建一个 Android 应用，Application name（应用名）取名为 RadioButton，Company Domain（公司域）取名为 Example.com
2	Minimum SDK 选择 API 18: Android 4.3 (Jelly Bean)
3	选择 Empty Activity
4	不需要修改 Activity Name，使用默认值，单击"Finish"按钮（完成）
5	在工程中找到 ras/layout 目录中的 activity_main.xml 文件，在其中添加一个 TextView（文本框）控件、一个 ImageButton（图片按钮）、一个 Button（普通按钮）控件和一个 RadioGroup（按钮组），在 RadioGroup（按钮组）中放置三个 RadioButton（单选按钮），按图 2.23 所示的运行效果位置进行摆放
6	将工程 res/mipmap 目录下 ic_launcher.png 图片文件复制、粘贴到 res/drawable 目录下
7	启动 Genymotion 模拟器，然后在 Android 工程中做如下代码修改

修改新建工程 res/layout 目录下的布局文件 activity_main.xml，修改代码如下：

```xml
<?xml version="1.0" encoding="utf-8"?>
<RelativeLayout xmlns:android="http://schemas.android.com/apk/res/android"
    xmlns:tools="http://schemas.android.com/tools" android:layout_width=
        "match_parent"
    android:layout_height="match_parent" android:paddingLeft="@dimen/
            activity_horizontal_margin"
    android:paddingRight="@dimen/activity_horizontal_margin"
    android:paddingTop="@dimen/activity_vertical_margin"
    android:paddingBottom="@dimen/activity_vertical_margin" tools:context=
            ".MainActivity">

    <TextView
        android:id="@+id/textView1"
        android:layout_width="wrap_content"
        android:layout_height="wrap_content"
        android:text="Example of Radio Button"
        android:layout_alignParentTop="true"
        android:layout_centerHorizontal="true"
        android:textSize="30dp" />

    <ImageButton
        android:layout_width="wrap_content"
        android:layout_height="wrap_content"
        android:id="@+id/imageButton"
        android:src="@drawable/ic_launcher"
        android:layout_below="@+id/textView1"
        android:layout_centerHorizontal="true" />
```

```xml
<Button
    android:layout_width="wrap_content"
    android:layout_height="wrap_content"
    android:id="@+id/button2"
    android:text="ClickMe"
    android:layout_alignParentBottom="true"
    android:layout_centerHorizontal="true" />

<RadioGroup

    android:layout_width="fill_parent"
    android:layout_height="fill_parent"
    android:layout_below="@+id/imageButton"
    android:id="@+id/radiogroup1"
    android:layout_centerHorizontal="true">

    <RadioButton
        android:layout_width="142dp"
        android:layout_height="wrap_content"
        android:text="JAVA"
        android:id="@+id/radioButton"
        android:textSize="25dp"
        android:textColor="@android:color/holo_red_light"
        android:checked="false"
        android:layout_gravity="center_horizontal" />

    <RadioButton
        android:layout_width="wrap_content"
        android:layout_height="wrap_content"
        android:text="ANDROID"
        android:id="@+id/radioButton2"
        android:layout_gravity="center_horizontal"
        android:checked="false"
        android:textColor="@android:color/holo_red_dark"
        android:textSize="25dp" />

    <RadioButton
        android:layout_width="136dp"
        android:layout_height="wrap_content"
        android:text="HTML"
        android:id="@+id/radioButton3"
        android:layout_gravity="center_horizontal"
        android:checked="false"
        android:textSize="25dp"
```

```
                    android:textColor="@android:color/holo_red_dark" />

        </RadioGroup>
</RelativeLayout>
```

修改新建工程 java/com.example.radiobutton 目录下的 java 文件 MainActivity.java,修改代码如下:

```java
package com.example.radiobutton;

import android.os.Bundle;
import android.support.v7.app.AppCompatActivity;
import android.view.View;
import android.widget.Button;
import android.widget.RadioButton;
import android.widget.RadioGroup;
import android.widget.Toast;

public class MainActivity extends AppCompatActivity {
    RadioGroup rg1;
    RadioButton rb1;
    Button b1;
    @Override
    protected void onCreate(Bundle savedInstanceState) {
        super.onCreate(savedInstanceState);
        setContentView(R.layout.activity_main);
        addListenerRadioButton();
    }
    private void addListenerRadioButton() {
        rg1 = (RadioGroup) findViewById(R.id.radiogroup1);
        b1 = (Button) findViewById(R.id.button2);
        b1.setOnClickListener(new View.OnClickListener() {
            @Override
            public void onClick(View v) {
                int selected = rg1.getCheckedRadioButtonId();
                rb1 = (RadioButton) findViewById(selected);
                Toast.makeText(MainActivity.this, rb1.getText(), Toast.
                        LENGTH_LONG).show();
            }
        });
    }
}
```

在 Genymotion 中运行,然后选中其中一个 RadioButton(单选按钮)选项,并单击 "CLICKME" 按钮,效果如图 2.23 所示。

图 2.23　RadioButton(单选按钮)与 RadioGroup(按钮组)实例运行效果

2.4.9　使用 ProgressDialog(进度对话框)类创建 ProgressBar(进度条)

当一个应用程序在后台执行时，前台界面不会有任何信息，这时用户不知道程序是否在执行以及执行进度等信息，因此需要使用进度条来提示程序执行的进度。例如，当用户从互联网上传或下载文件的时候，上传/下载进度条将显示下载进度信息。

在 Android 中，可以使用 ProgressDialog(进度对话框)类来创建 ProgressBar(进度条)控件。先用 ProgressDialog 类实例化一个对象来实现，语法如下：

```
ProgressDialog progress = new ProgressDialog(this);
```

然后再设置 ProgressDialog 类的属性，如下所示：

```
progress.setMessage("Downloading Music ! ");
progress.setProgressStyle(ProgressDialog.STYLE_HORIZONTAL);
progress.setIndeterminate(true);
```

● ProgressDialog 类的常用方法

ProgressDialog 的常用方法如表 2.33 所示。

表 2.33　ProgressDialog 的常用方法

方法	描述
getMax()	此方法将返回进度的最大值
incrementProgressBy(int diff)	此方法将设置增加的进度、每次推进的步伐
setIndeterminate(boolean indeterminate)	此方法将设置进度条是否为不确定模式
setMax(int max)	此方法将设置进度的最大值
setProgress(int value)	此方法将设置进度条的当前进度值
show(Context context, CharSequence title, CharSequence message)	这是一个静态方法，用于显示进度对话框

【例 2.16】 ProgressBar(进度条)实例

这个例子将使用 ProgressDialog(进度对话框)和 ProgressBar(进度条)控件来创建一个简单的 Android 应用,下面将通过表 2.34 中的步骤来实现。

表 2.34 ProgressDialog/ProgressBar 实例步骤

步骤	描述
1	打开 Android Studio 创建一个 Android 应用,Application name(应用名)取名为 ProgressBar,Company Domain(公司域)取名为 Example.com
2	Minimum SDK 选择 API 18: Android 4.3(Jelly Bean)
3	选择 Empty Activity
4	不需要修改 Activity Name,使用默认值,单击"Finish"按钮(完成)
5	在工程中找到 ras/layout 目录中的 activity_main.xml 文件,在其中添加一个 TextView(文本框)控件、一个 Button(普通按钮)控件,按图 2.24 所示的运行效果位置进行摆放
6	启动 Genymotion 模拟器,然后在 Android 工程中做如下代码修改

修改新建工程 res/layout 目录下的布局文件 activity_main.xml,修改代码如下:

```xml
<?xml version="1.0" encoding="utf-8"?>
<RelativeLayout xmlns:android="http://schemas.android.com/apk/res/android"
    xmlns:tools="http://schemas.android.com/tools" android:layout_width=
        "match_parent"
    android:layout_height="match_parent" android:paddingLeft="@dimen/
                    activity_horizontal_margin"
    android:paddingRight="@dimen/activity_horizontal_margin"
    android:paddingTop="@dimen/activity_vertical_margin"
    android:paddingBottom="@dimen/activity_vertical_margin" tools:context=
        ".MainActivity">

    <TextView
        android:layout_width="wrap_content"
        android:layout_height="wrap_content"
        android:id="@+id/textView"
        android:layout_alignParentTop="true"
        android:layout_centerHorizontal="true"
        android:textSize="30dp"
        android:text="Progress bar" />

    <Button
        android:layout_width="wrap_content"
        android:layout_height="wrap_content"
        android:text="Download"
        android:onClick="download"
        android:id="@+id/button2"
        android:layout_marginLeft="125dp"
        android:layout_marginStart="125dp"
        android:layout_centerVertical="true" />
</RelativeLayout>
```

修改新建工程 java/com.example.progressbar 目录下的 java 文件 MainActivity.java，修改代码如下：

```java
package com.example.progressbar;

import android.app.ProgressDialog;
import android.os.Bundle;
import android.support.v7.app.AppCompatActivity;
import android.view.View;
import android.widget.Button;

public class MainActivity extends AppCompatActivity {
    Button b1;
    private ProgressDialog progress;
    @Override
    protected void onCreate(Bundle savedInstanceState) {
        super.onCreate(savedInstanceState);
        setContentView(R.layout.activity_main);
        b1 = (Button) findViewById(R.id.button2);
    }
    public void download(View view){
        progress=new ProgressDialog(this);
        progress.setMessage("Downloading Music");
        progress.setProgressStyle(ProgressDialog.STYLE_HORIZONTAL);
        progress.setIndeterminate(true);
        progress.setProgress(0);
        progress.show();

        final int totalProgressTime = 100;
        final Thread t = new Thread() {
            @Override
            public void run() {
                int jumpTime = 0;

                while(jumpTime < totalProgressTime) {
                    try {
                        sleep(200);
                        jumpTime += 5;
                        progress.setProgress(jumpTime);
                    }
                    catch (InterruptedException e) {
                        // TODO Auto-generated catch block
                        e.printStackTrace();
                    }
                }
            }
        };
        t.start();
```

```
        }
    }
```

在 Genymotion 中运行程序的效果如图 2.24 所示。

当用户单击"DOWNLOAD"按钮后将弹出进度条对话框,程序运行的效果如图 2.25 所示。

图 2.24　ProgressDialog(进度对话框)/ProgressBar (进度条)实例运行效果 1

图 2.25　ProgressDialog(进度对话框)/ProgressBar (进度条)实例运行效果 2

2.4.10　Spinner(列表选择框)

Android 中提供的 Spinner(列表选择框)相当于在网页中常见的下拉列表框,用于供用户进行选择,从而方便用户。在用户使用电子邮件时,可以选择 Reply(回复)、Reply all(回复全部)和 Forward(转发)菜单项,如图 2.26 所示。

【例 2.17】　Spinner(列表选择框)实例

这个例子将使用 Spinner(列表选择框)控件来创建一个简单的 Android 应用,下面将通过表 2.35 中的步骤来实现。

图 2.26　Spinner(列表选择框)实例

表 2.35　Spinner(列表选择框)实例步骤

步骤	描述
1	打开 Android Studio 创建一个 Android 应用,Application name(应用名)取名为 Spinner,Company Domain(公司域)取名为 Example.com
2	Minimum SDK 选择 API 18: Android 4.3 (Jelly Bean)
3	选择 Empty Activity
4	不需要修改 Activity Name,使用默认值,单击"Finish"按钮(完成)
5	在工程中找到 ras/layout 目录中的 activity_main.xml 文件,在其中添加一个 TextView(文本框)控件、一个 Spinner(列表选择框),然后按图 2.27 所示的运行效果位置进行摆放
6	启动 Genymotion 模拟器,然后在 Android 工程中做如下代码修改

修改新建工程 res/layout 目录下的布局文件 activity_main.xml，修改代码如下：

```xml
<?xml version="1.0" encoding="utf-8"?>
<LinearLayout xmlns:android="http://schemas.android.com/apk/res/android"
    android:orientation="vertical"
    android:padding="10dip"
    android:layout_width="fill_parent"
    android:layout_height="wrap_content">

    <TextView
        android:layout_width="fill_parent"
        android:layout_height="wrap_content"
        android:layout_marginTop="10dip"
        android:text="Category:"
        android:layout_marginBottom="5dp"/>

    <Spinner
        android:id="@+id/spinner"
        android:layout_width="fill_parent"
        android:layout_height="wrap_content"
        />

</LinearLayout>
```

修改新建工程 java/com.example.spinner 目录下的 java 文件 MainActivity.java，修改代码如下：

```java
package com.example.spinner;

import android.app.Activity;
import android.os.Bundle;
import android.view.View;
import android.widget.AdapterView;
import android.widget.AdapterView.OnItemSelectedListener;
import android.widget.ArrayAdapter;
import android.widget.Spinner;
import android.widget.Toast;

import java.util.ArrayList;
import java.util.List;

public class MainActivity extends Activity implements OnItemSelectedListener {

    @Override
    protected void onCreate(Bundle savedInstanceState) {
        super.onCreate(savedInstanceState);
```

```java
        setContentView(R.layout.activity_main);
// Spinner element
        Spinner spinner = (Spinner) findViewById(R.id.spinner);

        // Spinner click listener
        spinner.setOnItemSelectedListener(this);

        // Spinner Drop down elements
        List<String> categories = new ArrayList<String>();
        categories.add("Automobile");
        categories.add("Business Services");
        categories.add("Computers");
        categories.add("Education");
        categories.add("Personal");
        categories.add("Travel");

        // Creating adapter for spinner
        ArrayAdapter<String> dataAdapter = new ArrayAdapter<String>(this,
            android.R.layout.simple_spinner_item, categories);

        // Drop down layout style - list view with radio button
        dataAdapter.setDropDownViewResource(android.R.layout.simple_
            spinner_dropdown_item);

        // attaching data adapter to spinner
        spinner.setAdapter(dataAdapter);
    }

    @Override
    public void onItemSelected(AdapterView<?> parent, View view, int
        position, long id) {
        // On selecting a spinner item
        String item = parent.getItemAtPosition(position).toString();

        // Showing selected spinner item
        Toast.makeText(parent.getContext(), "Selected: " + item, Toast.
            LENGTH_LONG).show();
    }
    public void onNothingSelected(AdapterView<?> arg0) {
        // TODO Auto-generated method stub
    }

}
```

在 Genymotion 中运行程序的效果如图 2.27 所示。

第 2 章 Android UI 设计

图 2.27　Spinner（列表选择框）实例运行效果

2.4.11　TimePicker（时间拾取器）

在 Android 中，TimePicker（时间拾取器）是用于选择一天中时间的控件，在 TimePicker 中可以设置小时、分钟和 AM/PM（上午/下午），如图 2.28 所示。

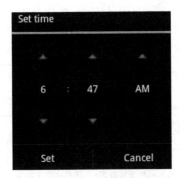

图 2.28　TimePicker（时间拾取器）实例

为了使用 TimePicker 类，首先需要在布局文件 activity.xml 中定义 TimePicker 组件，语法如下：

```
<TimePicker
    android:id="@+id/timePicker1"
    android:layout_width="wrap_content"
    android:layout_height="wrap_content" />
```

定义 TimePicker 组件后，需要在 Java 文件中实例化这个组件，实例化 TimePicker 的语法如下：

```
import android.widget.TimePicker;
```

```
private TimePicker timePicker1;
timePicker1 = (TimePicker) findViewById(R.id.timePicker1);
```

实例化组件后，可以通过 getCurrentHour()方法和 getCurrentMinute()方法获取当前系统小时和分钟数，语法如下：

```
int hour = timePicker1.getCurrentHour();
int min = timePicker1.getCurrentMinute();
```

● TimePicker 类的常用方法

TimePicker 类的常用方法如表 2.36 所示。

表 2.36 TimePicker 的常用方法

方法	描述
is24HourView()	此方法将返回当前 TimePicker 是否以 24 小时模式显示的状态
isEnabled()	此方法将返回当前 TimePicker 是否可用的状态
setCurrentHour(Integer currentHour)	此方法将设置 TimePicker 中显示的当前小时数
setCurrentMinute(Integer currentMinute)	此方法将设置 TimePicker 中显示的当前分钟数
setEnabled(boolean enabled)	此方法将设置 TimePicker 是否可用
setIs24HourView(Boolean is24HourView)	此方法将设置是否以 24 小时模式显示时间
setOnTimeChangedListener(TimePicker.OnTimeChangedListener onTimeChangedListener)	此方法将用于监听事件改变时触发的事件

【例 2.18】 TimePicker(时间拾取器)实例

这个例子将使用 TimePicker(时间拾取器)控件来创建一个简单的 Android 应用，下面将通过表 2.37 中的步骤来实现。

表 2.37 TimePicker(时间拾取器)实例步骤

步骤	描述
1	打开 Android Studio 创建一个 Android 应用，Application name(应用名)取名为 TimePicker，Company Domain(公司域)取名为 Example.com
2	Minimum SDK 选择 API 18: Android 4.3 (Jelly Bean)
3	选择 Empty Activity
4	不需要修改 Activity Name，使用默认值，单击"Finish"按钮(完成)
5	在工程中找到 ras/layout 目录中的 activity_main.xml 文件，在其中添加三个 TextView(文本框)控件、一个 Button(普通按钮)和一个 TimePicker(时间拾取器)，然后按图 2.29 所示的运行效果位置进行摆放
6	启动 Genymotion 模拟器，然后在 Android 工程中做如下代码修改

修改新建工程 res/layout 目录下的布局文件 activity_main.xml，修改代码如下：

```
<?xml version="1.0" encoding="utf-8"?>
<RelativeLayout xmlns:android="http://schemas.android.com/apk/res/android"
    xmlns:tools="http://schemas.android.com/tools" android:layout_width=
        "match_parent"
    android:layout_height="match_parent" android:paddingLeft="@dimen/
            activity_horizontal_margin"
    android:paddingRight="@dimen/activity_horizontal_margin"
    android:paddingTop="@dimen/activity_vertical_margin"
    android:paddingBottom="@dimen/activity_vertical_margin" tools:context=
        ".MainActivity">
```

```xml
<TextView
    android:id="@+id/textView2"
    android:layout_width="wrap_content"
    android:layout_height="wrap_content"
    android:layout_alignParentTop="true"
    android:layout_centerHorizontal="true"
    android:text="@string/time_pick"
    android:textAppearance="?android:attr/textAppearanceMedium" />

<Button
    android:id="@+id/set_button"
    android:layout_width="wrap_content"
    android:layout_height="wrap_content"
    android:layout_alignParentBottom="true"
    android:layout_centerHorizontal="true"
    android:layout_marginBottom="180dp"
    android:onClick="setTime"
    android:text="@string/time_save" />

<TimePicker
    android:id="@+id/timePicker1"
    android:layout_width="wrap_content"
    android:layout_height="wrap_content"
    android:layout_above="@+id/set_button"
    android:layout_centerHorizontal="true"
    android:layout_marginBottom="24dp" />

<TextView
    android:id="@+id/textView3"
    android:layout_width="wrap_content"
    android:layout_height="wrap_content"
    android:layout_alignLeft="@+id/timePicker1"
    android:layout_alignTop="@+id/set_button"
    android:layout_marginTop="67dp"
    android:text="@string/time_current"
    android:textAppearance="?android:attr/textAppearanceMedium" />

<TextView
    android:id="@+id/textView1"
    android:layout_width="wrap_content"
    android:layout_height="wrap_content"
    android:layout_below="@+id/textView3"
    android:layout_centerHorizontal="true"
    android:text="@string/time_selected"
    android:textAppearance="?android:attr/textAppearanceMedium" />
</RelativeLayout>
```

修改新建工程 res/values/目录下的数组文件 strings.xml，修改代码如下：

```xml
<resources>
    <string name="app_name">TimePicker</string>
    <string name="time_picker_example">Time Picker Example</string>
    <string name="time_pick">Pick the time and press save button</string>
    <string name="time_save">Save</string>
    <string name="time_selected"></string>
    <string name="time_current">The Time is:</string>
</resources>
```

修改新建工程 java/com.example.timepicker 目录下的 java 文件 MainActivity.java，修改代码如下：

```java
package com.example.timepicker;

import android.os.Bundle;
import android.support.v7.app.AppCompatActivity;
import android.view.View;
import android.widget.TextView;
import android.widget.TimePicker;

import java.util.Calendar;

public class MainActivity extends AppCompatActivity {
    private TimePicker timePicker1;
    private TextView time;
    private Calendar calendar;
    private String format = "";
    @Override
    protected void onCreate(Bundle savedInstanceState) {
        super.onCreate(savedInstanceState);
        setContentView(R.layout.activity_main);
        timePicker1 = (TimePicker) findViewById(R.id.timePicker1);
        time = (TextView) findViewById(R.id.textView1);
        calendar = Calendar.getInstance();

        int hour = calendar.get(Calendar.HOUR_OF_DAY);
        int min = calendar.get(Calendar.MINUTE);
        showTime(hour, min);
    }

    public void setTime(View view) {
        int hour = timePicker1.getCurrentHour();
        int min = timePicker1.getCurrentMinute();
        showTime(hour, min);
    }

    public void showTime(int hour, int min) {
        if (hour == 0) {
```

```
            hour += 12;
            format = "AM";
        }
        else if (hour == 12) {
            format = "PM";
        } else if (hour > 12) {
            hour -= 12;
            format = "PM";
        } else {
            format = "AM";
        }
        time.setText(new StringBuilder().append(hour).append(" : ").append(min)
                .append(" ").append(format));
    }
}
```

在 Genymotion 中运行程序的效果如图 2.29 所示。

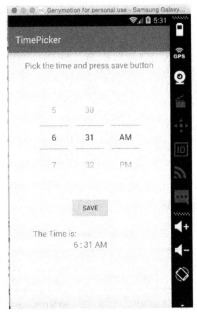

图 2.29　TimePicker(时间拾取器)实例运行效果

2.5　DatePicker(日期拾取器)与 DatePickerDialog (日期拾取器对话框)

在 Android 中，DatePicker(日期拾取器)是让用户在界面中选择日期的控件。DatePicker 由年、月、日三个部分组成。Android 提供 DatePicker 控件和调用 DatePickerDialog(日期拾取器对话框)类两种方式，以实现日期选取功能。

在本节中，将通过调用 DatePickerDialog 类的方式来实现日期选取功能。要在屏幕上显示 DatePickerDialog，需要通过 showDialog(id_of_dialog) 来调用，其语法如下：

```
showDialog(999);  //其中 999 为 DatePickerDialog id
```

调用 showDialog 方法后，onCreateDialog 方法将自动被调用，所以需要重写 onCreateDialog 方法，其语法如下：

```java
@Override
protected Dialog onCreateDialog(int id) {
    // TODO Auto-generated method stub
    if (id == 999) {
        return new DatePickerDialog(this, myDateListener, year, month, day);
    }
    return null;
}
```

最后一步，需要注册 DatePickerDialog 监听器且重写 onDateSet 方法。onDateSet 方法包含更新后的年、月、日数据。DatePickerDialog 监听器与 onDateSet 方法的语法如下所示：

```java
private DatePickerDialog.OnDateSetListener myDateListener = new
        DatePickerDialog.OnDateSetListener() {
    @Override
    public void onDateSet(DatePicker arg0, int arg1, int arg2, int arg3) {
        // arg1 = year
        // arg2 = month
        // arg3 = day
    }
};
```

- DatePicker 的常用方法

DatePicker 的常用方法如表 2.38 所示。

表 2.38　DatePicker 的常用方法

方法	描述
getDayOfMonth()	此方法将返回当前 DatePicker 中的天数值
getMonth()	此方法将返回当前 DatePicker 中的月份值
getYear()	此方法将返回当前 DatePicker 中的年
setMaxDate(long maxDate)	此方法将设置 DatePicker 所支持的最大日期数
setMinDate(long minDate)	此方法将设置 DatePicker 所支持的最小日期数
setSpinnersShown(boolean shown)	此方法将设置 DatePicker 是否显示下拉选择框
getCalendarView()	此方法将返回 calendar view
getFirstDayOfWeek()	此方法将返回一周的第一天是星期几
updateDate(int year, int month, int dayOfMonth)	此方法将通过参数更新 DatePicker 的现有时间

【例 2.19】　DatePicker（日期拾取器）实例

这个例子将使用 DatePicker（日期拾取器）控件来创建一个简单的 Android 应用，下面将通过表 2.39 中的步骤来实现。

第 2 章 Android UI 设计

表 2.39 DatePicker(日期拾取器)实例步骤

步骤	描述
1	打开 Android Studio 创建一个 Android 应用，Application name(应用名)取名为 DatePicker, Company Domain(公司域)取名为 Example.com
2	Minimum SDK 选择 API 18: Android 4.3(Jelly Bean)
3	选择 Empty Activity
4	不需要修改 Activity Name，使用默认值，单击 "Finish" 按钮(完成)
5	在工程中找到 ras/layout 目录中的 activity_main.xml 文件，在其中添加三个 TextView(文本框)控件、一个 Button(普通按钮)，然后按图 2.30 所示的运行效果位置进行摆放
6	启动 Genymotion 模拟器，然后在 Android 工程中做如下代码修改

修改新建工程 res/layout 目录下的布局文件 activity_main.xml，修改代码如下：

```xml
<?xml version="1.0" encoding="utf-8"?>
<RelativeLayout xmlns:android="http://schemas.android.com/apk/res/android"
    xmlns:tools="http://schemas.android.com/tools" android:layout_width=
            "match_parent"
    android:layout_height="match_parent" android:paddingLeft="@dimen/
                    activity_horizontal_margin"
    android:paddingRight="@dimen/activity_horizontal_margin"
    android:paddingTop="@dimen/activity_vertical_margin"
    android:paddingBottom="@dimen/activity_vertical_margin" tools:context=
                    ".MainActivity">

    <Button
        android:id="@+id/button1"
        android:layout_width="wrap_content"
        android:layout_height="wrap_content"
        android:layout_alignParentTop="true"
        android:layout_centerHorizontal="true"
        android:layout_marginTop="70dp"
        android:onClick="setDate"
        android:text="@string/date_button_set" />

    <TextView
        android:id="@+id/textView1"
        android:layout_width="wrap_content"
        android:layout_height="wrap_content"
        android:layout_alignParentTop="true"
        android:layout_centerHorizontal="true"
        android:layout_marginTop="24dp"
        android:text="@string/date_label_set"
        android:textAppearance="?android:attr/textAppearanceMedium" />

    <TextView
        android:id="@+id/textView2"
        android:layout_width="wrap_content"
        android:layout_height="wrap_content"
```

```xml
            android:layout_below="@+id/button1"
            android:layout_marginTop="66dp"
            android:layout_toLeftOf="@+id/button1"
            android:text="@string/date_view_set"
            android:textAppearance="?android:attr/textAppearanceMedium" />

    <TextView
            android:id="@+id/textView3"
            android:layout_width="wrap_content"
            android:layout_height="wrap_content"
            android:layout_alignRight="@+id/button1"
            android:layout_below="@+id/textView2"
            android:layout_marginTop="72dp"
            android:text="@string/date_selected"
            android:textAppearance="?android:attr/textAppearanceMedium" />
</RelativeLayout>
```

修改新建工程 res/values/目录下的数组文件 strings.xml，修改代码如下：

```xml
<resources>
    <string name="app_name">DatePicker</string>
    <string name="hello_world">Hello world!</string>
    <string name="date_label_set">Press the button to set the date</string>
    <string name="date_button_set">Set Date</string>
    <string name="date_view_set">The Date is: </string>
    <string name="date_selected"></string>
</resources>
```

修改新建工程 java/com.example.datapicker 目录下的 java 文件 MainActivity.java，修改代码如下：

```java
package com.example.datepicker;

import android.app.DatePickerDialog;
import android.app.Dialog;
import android.os.Bundle;
import android.support.v7.app.AppCompatActivity;
import android.view.View;
import android.widget.DatePicker;
import android.widget.TextView;
import android.widget.Toast;

import java.util.Calendar;

public class MainActivity extends AppCompatActivity {
    private DatePicker datePicker;
    private Calendar calendar;
    private TextView dateView;
    private int year, month, day;
```

```java
@Override
protected void onCreate(Bundle savedInstanceState) {
    super.onCreate(savedInstanceState);
    setContentView(R.layout.activity_main);
    dateView = (TextView) findViewById(R.id.textView3);
    calendar = Calendar.getInstance();
    year = calendar.get(Calendar.YEAR);

    month = calendar.get(Calendar.MONTH);
    day = calendar.get(Calendar.DAY_OF_MONTH);
    showDate(year, month + 1, day);
}
@SuppressWarnings("deprecation")
public void setDate(View view) {
    showDialog(999);
    Toast.makeText(getApplicationContext(), "ca", Toast.LENGTH_SHORT)
        .show();
}


@Override
protected Dialog onCreateDialog(int id) {
    // TODO Auto-generated method stub
    if (id == 999) {
        return new DatePickerDialog(this, myDateListener, year, month, day);
    }
    return null;
}

private DatePickerDialog.OnDateSetListener myDateListener = new
        DatePickerDialog.OnDateSetListener() {
    @Override
    public void onDateSet(DatePicker arg0, int arg1, int arg2, int arg3) {
        // TODO Auto-generated method stub
        // arg1 = year
        // arg2 = month
        // arg3 = day
        showDate(arg1, arg2+1, arg3);
    }
};

private void showDate(int year, int month, int day) {
    dateView.setText(new StringBuilder().append(day).append("/")
        .append(month).append("/").append(year));
}
}
```

在 Genymotion 中运行程序的效果如图 2.30 所示。

当用户单击"SET DATE"按钮后,将弹出日期选取对话框,程序运行的效果如图 2.31 所示。

96　Android 应用程序开发教程——Android Studio 版

图 2.30　DatePicker（日期拾取器）实例运行效果 1

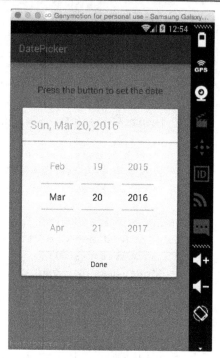

图 2.31　DatePicker（日期拾取器）实例运行效果 2

第 3 章 基本程序单元 Activity

在 Android 中，Activity（活动）表示一个单一屏幕上的用户界面。例如，电子邮件应用程序主界面可能是一个 Activity，显示新的电子邮件列表则是另一个 Activity，撰写电子邮件又是其他的 Activity。如果应用程序有一个以上的 Activity，则应该在 AndroidManifest.xml 配置文件中将其中一个 Activity 标记为启动 Activity 以开启应用程序。

如果用户曾经使用过 C、C++和 Java，就会知道一般程序是从 main()函数开始运行的。相似地，Android 系统是从 MainActivity 的 onCreate()方法调用开始启动程序的。在图 3.1 中，用矩形方框表示可以被回调的方法，椭圆形表示 Acitivity 的状态。从该图中可以看出，在一个 Activity 的生命周期中有表 3.1 中的方法会被系统调用。

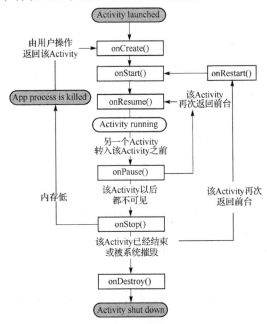

图 3.1 Activity 的生命周期及回调方法

Activity 生命周期中的回调方法如表 3.1 所示。

表 3.1 Activity 生命周期中的回调方法

回调方法	描述
onCreate()	在创建 Activity 时被调用。该方法是最常见的方法，在 Android Studio 创建项目时，会自动创建一个 MainActivity，在该 MainActivity 中，默认重写了 onCreate(Bundle savedInstanceState)方法，用于对该 Activity 进行初始化
onStart()	启动 Activity 时被回调，也就是当一个 Activity 变为显示时被回调
onResume()	当 Activity 由暂停状态恢复为活动状态时调用。调用该方法后，该 Activity 位于 Activity 堆栈的栈顶。该方法总是在 onPause()方法以后执行
onPause()	暂停 Activity 时被调用。该方法需要被非常快地执行，应用直到该方法执行完毕后，下一个 Activity 才能被恢复。该方法通常用于保存数据。比如，当用户玩游戏时，突然来了一个电话，这时在该方法中保存游戏状态的数据
onStop()	停止 Activity 时被回调
onDestroy()	销毁 Activity 时被回调
onRestart()	重新启动 Activity 时被回调，该方法总是在 onStart()方法后被执行

【例 3.1】 Activity（活动）实例

这个 Activity（活动）的例子将通过表 3.2 中的步骤来修改 HelloWorld 例子实现。

表 3.2　Activity（活动）实例步骤

步骤	描述
1	打开 Android Studio 创建一个 Android 应用，Application name（应用名）取名为 Activity，Company Domain（公司域）取名为 Example.com
2	Minimum SDK 选择 API 18: Android 4.3 (Jelly Bean)
3	选择 Empty Activity
4	不需要修改 Activity Name，使用默认值，单击"Finish"按钮（完成）
5	启动 Genymotion 模拟器，然后在 Android 工程中做如下代码修改

修改工程的 java/com.example.activtiy 目录下 java 文件 MainActivity.java，修改代码如下：

```java
package com.example.activity;

import android.app.Activity;
import android.os.Bundle;
import android.util.Log;

public class MainActivity extends Activity {

    String msg = "Android : ";

    /** Called when the activity is first created. */
    @Override
    public void onCreate(Bundle savedInstanceState) {
        super.onCreate(savedInstanceState);
        setContentView(R.layout.activity_main);
        Log.d(msg, "The onCreate() event");
    }

    /** Called when the activity is about to become visible. */
    @Override
    protected void onStart() {
        super.onStart();
        Log.d(msg, "The onStart() event");
    }

    /** Called when the activity has become visible. */
    @Override
    protected void onResume() {
        super.onResume();
        Log.d(msg, "The onResume() event");
    }

    /** Called when another activity is taking focus. */
    @Override
```

```java
    protected void onPause() {
        super.onPause();
        Log.d(msg, "The onPause() event");
    }

    /** Called when the activity is no longer visible. */
    @Override
    protected void onStop() {
        super.onStop();
        Log.d(msg, "The onStop() event");
    }

    /** Called just before the activity is destroyed. */
    @Override
    public void onDestroy() {
        super.onDestroy();
        Log.d(msg, "The onDestroy() event");
    }
}
```

Activity（活动）通过 setContentView 方法加载工程 res/layout 目录下的 XML 文件从而调用 UI 组件。setContentView 加载 activity_main.xml 文件语句如下：

```java
setContentView(R.layout.activity_main);
```

Android 应用程序中可以有一个或多个 Activity（活动），但每个 Activity（活动）都需要在 AndroidManifest.xml 配置文件中进行声明。同时，主 Activity（活动）还需要用<intent-filter>标签标明为 MAIN 动作和 LAUNCHER 类，语法如下所示：

```xml
<?xml version="1.0" encoding="utf-8"?>
<manifest xmlns:android="http://schemas.android.com/apk/res/android"
    package="com.example.activity" >

    <application
        android:allowBackup="true"
        android:icon="@mipmap/ic_launcher"
        android:label="@string/app_name"
        android:supportsRtl="true"
        android:theme="@style/AppTheme" >
        <activity android:name=".MainActivity" >
            <intent-filter>
                <action android:name="android.intent.action.MAIN" />

                <category android:name="android.intent.category.LAUNCHER" />
            </intent-filter>
        </activity>
    </application>

</manifest>
```

注意：如果在 AndroidManifest.xml 配置文件中没有任何一个 Activity（活动）标注了 MAIN 动作和 LAUNCHER 类，那么用户在手机屏幕上将看不到应用程序的 APP 图标。

运行例 3.1 程序，在 Genymotion 中的运行效果如图 3.2 所示。

图 3.2　Activity（活动）实例运行效果

返回 Android Studio 界面，如图 3.3 所示，查看 Logcat 窗口中 Activity（活动）调用 Log.d() 方法的运行结果。

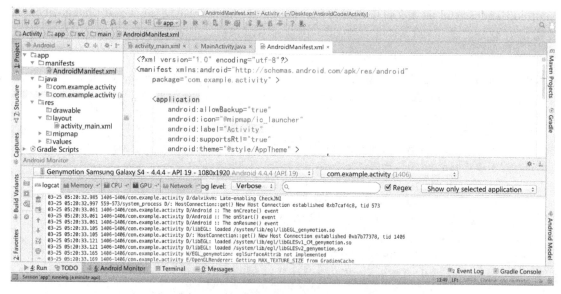

图 3.3　在 Android Studio 中查看 Logcat

在 Logcat 中找到 Log.d() 方法的运行结果：

```
1406-1406/com.example.activity D/Android :: The onCreate() event
1406-1406/com.example.activity D/Android :: The onStart() event
1406-1406/com.example.activity D/Android :: The onResume() event
```

然后，尝试单击 Genymotion 模拟器中的 ⌂Home 键，在 Logcat 中找到 Log.d()方法的运行结果：

```
1406-1406/com.example.activity D/Android :: The onPause() event
1406-1406/com.example.activity D/Android :: The onStop() event
```

再尝试单击 Genymotion 模拟器中的 ☰Menu 键，在 Logcat 中找到 Log.d()方法的运行结果：

```
1406-1406/com.example.activity D/Android :: The onStart() event
1406-1406/com.example.activity D/Android :: The onResume() event
```

最后，尝试单击 Genymotion 模拟器中的 ↶Back 键，在 Logcat 中找到 Log.d()方法的运行结果：

```
1406-1406/com.example.activity D/Android :: The onPause() event
1406-1406/com.example.activity D/Android :: The onStop() event
1406-1406/com.example.activity D/Android :: The onDestroy() event
```

第 4 章　Android 应用核心 Intent 和 Filters

在 Android 中，Intent(意图)是一个将要执行操作的抽象描述。Android 的 3 个核心组件 Activity(活动)、Service(服务)和 BroadcastReceiver(广播接收器)都需要使用 Intent 来进行激活。Intent 用于相同或者不同应用程序组件间的后期运行时绑定。

对于不同的组件，Android 系统提供了不同的 Intent 启动方法，如表 4.1 所示。

表 4.1　不同组件的启动 Intent 方法

组件类型	启动方法
Activity	startActivity(Intent intent)
	startActivityForResult(Intent intent, int requestCode)
Service	ComponentName StartService(Intent service)
	Boolean bindService(Intent service, ServiceConnection conn, int flags)
BroadcastReceiver	sendBroadcast(Intent intent)
	sendBroadcast(Intent intent, String receiverPermission)
	sendOrderedBroadcast(Intent intent, String receiverPermission, BroadcastReceiver resultReceiver, Handler scheduler, int initialCode, String initialData, Bundle initialExtras)
	sendOrderedBroadcast(Intent intent, String receiverPermission)
	sendStickyBroadcast(Intent intent)
	sendStickyOrderedBroadcst(Intent intent, BroadcstReceiver resultReceiver, Handler scheduler, int initialCode, String initialData, Bundle initialExtras)

例如，你有一个 Activity 需要调用 Android 系统中的浏览器打开一个 URL。为了达到这个目的，Activity 将发送 ACTION_WEB_SEARCH Intent 意图到 Android Intent Resolver(意图解析器)，Android Intent Resolver 通过解析 Activity 列表选择一个最合适的 Activity 来执行这个意图。这里，Android Intent Resolver 将选择 Web Browser Activity(网页浏览器活动)来打开 URL。

```
String q = "tutorialspoint";
Intent intent = new Intent(Intent.ACTION_WEB_SEARCH );
intent.putExtra(SearchManager.QUERY, q);
startActivity(intent);
```

上面的代码表示 Android Intent Resolver 将打开搜索引擎搜索"tutorialspoint"关键字。

4.1　Intent 的各属性

Intent 对象中包含了接收该 Intent 的组件感兴趣的信息(如执行的操作和操作的数据)和 Android 系统感兴趣的信息(如处理该 Intent 的组件的类别和任何启动目标 Activity 的说明)。Intent 对象大致包含 Component(组件)、Action(动作)、Category(类别)、Data(数据)、Type(类型)、Extra(额外)和 Flag(标记)等属性，下面详细介绍 Intent 对象各属性的作用。

4.1.1 Component(组件)

Intent 的 Component(组件)内容需要接受一个 ComponentName(组件名称)属性，ComponentName 属性包含如表 4.2 所示的几个构造器。

表 4.2 ComponentName 属性的构造器

构造器	描述
ComponentName(String pkg, String cls)	创建 pkg 所在包下的 cls 类所对应的组件
ComponentName(Context pkg, String cls)	创建 pkg 所对应的包下的 cls 类所对应的组件
ComponentName(context pkg, Class<?> cls)	创建 pkg 所对应的包下的 cls 类所对应的组件

上面的构造器的本质就是一个，这说明创建一个 ComponentName 需要指定包名和类名(可唯一地确定的组件类)，这样应用程序就可根据给定的组件类去启动特定的组件。

除此之外，Intent(意图)还包含如表 4.3 所示的三个方法。

表 4.3 设置 Intent 启动组件类的三种方法

方法	描述
setClass(Context packageContext, Class<?> cls)	设置该 Intent 将要启动的组件的类
setClassName(Context packageContext, String className)	设置该 Intent 将要启动的组件对应的类名
setClassName(String packageName, String className)	设置该 Intent 将要启动的组件对应的类名

4.1.2 Action(动作)

Action(动作)是一个字符串，用来表示要完成的一个抽象动作。对这个动作具体由哪个组件(或许是 Activity、或许是 BroadcastReceiver)来完成，Action 这个字符串本身并不管。需要指出的是，一个 Intent 对象最多只能包括一个 Action 属性，Android 定义了一系列动作常量，其目标组件包括 Activity 和 BroadcastReceiver 两类。下面分别介绍这两类动作。

1. 标准 Activity 动作

表 4.4 中列出了当前 Intent 类中定义的用于启动 Activity 的标准动作，其中最常用的是 ACTION_MAIN 和 ACTION_EDIT。

表 4.4 Intent 类中定义的用于启动 Activity 的标准动作

Action 常量	对应字符串	描述
ACTION_MAIN	Android.intent.action.MAIN	应用程序入口
ACTION_VIEW	Android.intent.action.VIEW	显示指定数据
ACTION_ATTACH_DATA	Android.intent.action.ATTACH_DATA	指定某块数据将被附加到其他地方
ACTION_EDIT	Android.intent.action.EDIT	编辑指定数据
ACTION_PICK	Android.intent.action.PICK	从列表中选择某项，并返回所选的数据
ACTION_CHOOSER	Android.intent.action.CHOOSER	显示一个 Activity 选择器
ACTION_GET_CONTENT	Android.intent.action.GET_CONTENT	让用户选择数据，并返回所选数据
ACTION_DIAL	Android.intent.action.DIAL	显示拨号面板
ACTION_CALL	Android.intent.action.CALL	直接向指定用户打电话
ACTION_SEND	Android.intent.action.SEND	向其他人发送数据
ACTION_SENDTO	Android.intent.action.SENDTO	向其他人发送信息
ACTION_ANSWER	Android.intent.action.ANSWER	应答电话

Action 常量	对应字符串	描述
ACTION_INSERT	Android.intent.action.INSERT	插入数据
ACTION_DELETE	Android.intent.action.DELETE	删除数据
ACTION_RUN	Android.intent.action.RUN	运行数据
ACTION_SYNC	Android.intent.action.SYNC	执行数据同步
ACTION_PICK_ACTIVITY	Android.intent.action.ACTIVITY	用于选择 Activity
ACTION_SEARCH	Android.intent.action.SEARCH	执行搜索
ACTION_WEB_SEARCH	Android.intent.action.WEB_SEARCH	执行 Web 搜索
ACTION_FACTORY_TEST	Android.intent.action.FACTORY_TEST	工厂测试的入口

2．标准接收广播动作

表 4.5 中列出当前 Intent 类中定义的用于接收广播的标准动作。

表 4.5　Intent 类中定义的用于接收广播的标准动作

Action 常量	对应字符串	描述
ACTION_TIME_TICK	Android.intent.action.TIME_TICK	每分钟通知一次当前时间
ACTION_TIME_CHANGED	Android.intent.action.TIME_CHANGED	通知时间被修改
ACTION_TIMEZONE_CHANGED	Android.intent.action.TIMEZONE_CHANGED	通知时区被修改
ACTION_BOOT_COMPLETED	Android.intent.action.BOOT_COMPLETED	在系统启动完成后发出一次通知
ACTION_PACKAGE_ADDED	Android.intent.action.PACKAGE_ADDED	通知新应用程序包已经安装到设备上
ACTION_PACKAGE_CHANGED	Android.intent.action.PACKAGE_CHANGED	通知已经安装的应用程序包已经被修改
ACTION_PACKAGE_REMOVED	Android.intent.action.PACKAGE_REMOVED	通知从设备中删除应用程序包
ACTION_PACKAGE_RESTARTED	Android.intent.action._PACKAGE_RESTARTED	通知用户重启应用程序包，其所有进程都被关闭
ACTION_PACKAGE_DATA_CLEARED	Android.intent.action.PACKAGE_DATA_CLEARED	通知用户清空应用程序包中的数据
ACTION_UID_REMOVED	Android.intent.action.UID_REMOVED	通知从系统中删除用户 ID 值
ACTION_BATTERY_CHANGED	Android.intent.actoin.BATTERY_CHANGED	包含充电状态、等级和其他电池信息的广播
ACTION_POWER_CONNECTED	Android.intent.action.POWER_CONNECTED	通知设备已经连接外置电源
ACTION_POWER_DISCONNECTED	Android.intent.action.POWER_DISCONNECTED	通知设备已经移除外置电源
ACTION_SHUTDOWN	Android.intent.action.SHUTDOWN	通知设备已经关闭

注意：表 4.4 和表 4.5 所列出的都只是部分较为常用的 Action 常量，关于 Intent 所提供的全部 Action 常量，应参考 Android 官方 API 文档中关于 Intent 的说明。

4.1.3　Category（类别）

Category（类别）也是一个字符串，它用于为 Aciton 增加额外的附加类别信息。程序可调用 Intent 的 addCategory（string str）方法将为 Intent 添加 Category，removeCategory（）方法删除上次增加的 Category，getCategory（）方法获得当前对象中包含的全部 Category。

在 Intent 对象中可以添加任意多个 Category。与 Aciton 类似，在 Intent 类中也预定义了一些类别常量，常用的标准 Category 常量以及对应的字符串如表 4.6 所示。

表 4.6 常用的标准 Category 常量以及对应的字符串

Category(类别)常量	对应的字符串	简单说明
CATEGORY_DEFAULT	android.intent.category.DEFAULT	默认的 Category
CATEGORY_BROWSABLE	android.intent.category.BROWSABLE	指定该 Activity 能被浏览器安全调用
CATEGORY_TAB	android.intent.category.TAB	指定 Activity 作为 TabActivity 的 Tab 页
CATEGORY_LAUNCHER	android.intent.category.LAUNCHER	Activity 显示在顶级程序列表中
CATEGORY_INFO	android.intent.category.INFO	用于提供包信息
CATEGORY_HOME	android.intent.category.HOME	设置该 Activity 随系统启动而运行
CATEGORY_PREFERENCE	android.intent.category.PREFERENCE	该 Activity 是参数面板
CATEGORY_TEST	android.intent.category.TEST	该 Activity 是一个测试
CATEGORY_CAR_DOCK	android.intent.category.CAR_DOCK	指定手机被插入汽车硬件时运行该 Activity
CATEGORY_DESK_DOCK	android.intent.category.DESK_DOCK	指定手机被插入桌面硬件时运行该 Activity
CATEGORY_CAR_MODE	android.intent.category.CAR_MODE	设置该 Activity 可在车载环境下使用

注意：表 4.6 列出的都只是部分较为常用的 Category 常量，关于 Intent 所提供的全部 Category 常量，应参考 Android 官方 API 文档中关于 Intent 的说明。

4.1.4 Data(数据)和 Type(类型)

Data(数据)通常用于向 Action 属性提供操作的数据。Data 接受一个 URI 对象，一个 URI 对象通常通过如下形式的字符串来表示：

```
content://com.android.contacts/contacts/1
tel:123
```

URI 字符串通常是如下格式：

```
scheme://host:port/path
```

例如上面给出的 content://com.android.contacts/contacts/1，其中 content 是 scheme 部分，com.android.contacts 是 host 部分，port 部分被省略，/contacts/1 是 path 部分。

Type(类型)用于指定该 Data 所指定 URI 对应的 MIME 类型，这种 MIME 类型可以是任何自定义的 MIME 类型，只要符合 abc/xyz 格式的字符串即可。

Data 和 Type 的关系比较微妙，这两个属性会相互覆盖，例如：
- 如果为 Intent 先设置 Data，后设置 Type，那么 Type 属性将会覆盖 Data 属性。
- 如果为 Intent 先设置 Type，后设置 Data，那么 Data 属性将会覆盖 Type 属性。
- 如果希望 Intent 既有 Data 属性，也有 Type 属性，则需要调用 Intent 的 setDataAndType() 方法。

在 AndroidManifest.xml 配置文件中为组件声明 Data、Type 属性都通过<data.../>标签，<data.../>标签的格式如下：

```xml
<data android:mimeType=" "
    android:scheme=" "
    android:host=" "
    android:port=" "
    android:path=" "
    android:pathPrefix=" "
    android:pathPattern=" " />
```

上面的<data.../>标签支持如下属性。
- mimeType：用于声明该组件所能匹配的 Intent 的 Type 属性。
- scheme：用于声明该组件所能匹配的 Intent 的 scheme 部分。
- host：用于声明该组件所能匹配的 Intent 的 Data 属性的 host 部分。
- port：用于声明该组件所能匹配的 Intent 的 Data 属性的 port 部分。
- path：用于声明该组件所能匹配的 Intent 的 Data 属性的 path 部分。
- pathPrefix：用于声明该组件所能匹配的 Intent 的 Data 属性的 path 前缀。
- pathPattern：用于声明该组件所能匹配的 Intent 的 Data 属性的 path 字符串模板。

一旦为 Intent 同时指定了 Action 和 Data 属性，那么 Android 将可根据指定的数据类型来启动特定的应用程序，并对指定数据执行相应的操作。

下面是几个 Action 属性和 Data 属性组合例子。
- ACTION_VEW content://com.android.contacts/contacts/1：显示标记为 1 的联系人的信息。
- ACTION_EDIT content://com.android.contacts/contacts/1：编辑标识为 1 的联系人的信息。
- ACTION_DIAL content://com.android.contacts/contacts/1：显示给标识为 1 的联系人拨号的界面。
- ACTION_VEW tel:123：显示向指定号码 123 拨号的界面。
- ACTION_DIAL tel:123：显示向指定号码 123 拨号的界面。
- ACTION_VEW content://contacts/people/：显示所有联系人列表的信息。

4.1.5 Extra（额外）

Intent 的 Extra（额外）属性通常用于在多个 Activity 之间进行数据交换，Intent 的 Extra 属性值应该是一个 Bundle 对象，Bundle 对象是一个 Map 数据结构对象，它可以存入多组 key-value 对，Extra 可以通过 putExtras()方法和 getExtras()方法设置和读取 Bundle。这样，就可以通过 Intent 在不同 Activity 之间进行数据交换了。

4.1.6 Flag（标记）

Intent 的 Flag（标记）属性用于为该 Intent 添加一些额外的控制标记，Intent 可调用 addFlags()方法来为 Intent 添加控制标记。Intent 中常用的 Flag 如下：

表 4.7　Intent 中常用的 Flag

Flag（标记）	描述
FLAG_ACTIVITY_BROUGHT_TO_FRONT	如果通过该标记启动的 Activity 已经存在，则下次在启动时，只是将该 Activity 带到前台
FLAG_ACTIVITY_CLEAR_TOP	该标记相当于加载模式中的 singleTask，通过这种 Flag 启动的 Activity 将会把要启动的 Activity 之上的 Activity 全部弹出 Activity 栈
FLAG_ACTIVITY_NEW_TASK	默认的启动标记，该标记控制重新创建一个新的 Activity
FLAG_ACTIVITY_NO_ANIMATION	该标记会控制启动 Activity 时不使用过渡动画
FLAG_ACTIVITY_NO_HISTORY	该标记控制被启动的 Activity 将不会保留在 Activity 栈中
FLAG_ACTIVITY_REORDER_TO_FRONT	该标记控制如果当前已有该 Activity，则直接将该 Activity 带到前台
FLAG_ACTIVITY_SINGLE_TOP	该标记相当于加载模式中的 singleTop 模式

4.2 Intents 的类型

在 Android 中,Intent 分为显式 Intent 和隐式 Intent,如图 4.1 所示。

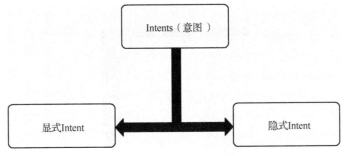

图 4.1 Intent 的分类

4.2.1 显式 Intent

显式 Intent 是指定了 Component(组件)属性的 Intent,它已经明确了将启动哪个组件。例如,你打算从一个 Activity 中通过单击按钮启动另一个 Activity,如图 4.2 所示。

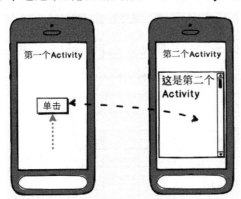

图 4.2 显式 Intent 举例

通过如下代码就能实现:

```
// Explicit Intent by specifying its class name
Intent i = new Intent(FirstActivity.this, SecondActivity.class);
i.putExtra("Key1", "ABC");
i.putExtra("Key2", "123");
// Starts TargetActivity
startActivity(i);
```

4.2.2 隐式 Intent

隐式 Intent 是没有指定 Component(组件)属性的 Intent。由于隐式 Intent 没有明确指定要启动哪个组件,应用程序将会根据 Intent 指定的规则去启动符合条件的组件,但具体是哪个组件则不确定(系统自动选择适合的组件执行 Intent)。例如,运行下面的代码:

```
Intent read1=new Intent();
read1.setAction(android.content.Intent.ACTION_VIEW);
read1.setData(ContactsContract.Contacts.CONTENT_URI);
startActivity(read1);
```

则得到如图 4.3 所示的运行效果。

图 4.3 隐式 Intent 举例

目标组件可以通过 getExtras()方法接收源组件传递过来的 Extra(额外)信息，代码如下：

```
// Get bundle object at appropriate place in your code
Bundle extras = getIntent().getExtras();

// Extract data using passed keys
String value1 = extras.getString("Key1");
String value2 = extras.getString("Key2");
```

【例 4.1】 Intent(意图)实例

这个例子将展示 Intent(意图)在 Android 中的应用。Intent(意图)1 实例步骤如表 4.8 所示。

表 4.8 Intent(意图)1 实例步骤

步骤	描述
1	打开 Android Studio 创建一个 Android 应用，Application name(应用名)取名为 Intent1，Company Domain(公司域)取名为 Example.com
2	Minimum SDK 选择 API 18: Android 4.3 (Jelly Bean)
3	选择 Empty Activity
4	不需要修改 Activity Name，使用默认值，单击"Finish"按钮(完成)
5	在工程中找到 ras/layout 目录中的 activity_main.xml 文件，在其中添加一个 TextView(文本框)控件、一个 ImageButton(图片按钮)和两个 Button(普通按钮)控件(START BROWSER 和 START PHONE)，按图 4.4 所示的运行效果位置进行摆放
6	将工程 res/mipmap 目录下的 ic_launcher.png 图片文件复制、粘贴到 res/drawable 目录下
7	修改 java/com.example.intent1 目录下 java 文件 MainActivity.java 代码，为 START BROWSER 和 START PHONE 两个按钮添加单击事件监听器
8	启动 Genymotion 模拟器，然后在 Android 工程中做如下代码修改

修改新建工程 res/layout 目录下的布局文件 activity_main.xml，修改代码如下：

```xml
<?xml version="1.0" encoding="utf-8"?>
<RelativeLayout xmlns:android="http://schemas.android.com/
    apk/res/android"
    xmlns:tools="http://schemas.android.com/tools" android:layout_width=
        "match_parent"
    android:layout_height="match_parent" android:paddingLeft="@dimen/
        activity_horizontal_margin"
    android:paddingRight="@dimen/activity_horizontal_margin"
    android:paddingTop="@dimen/activity_vertical_margin"
    android:paddingBottom="@dimen/activity_vertical_margin" tools:context=
        ".MainActivity">

    <TextView
        android:id="@+id/textView1"
        android:layout_width="wrap_content"
        android:layout_height="wrap_content"
        android:text="Intent Example"
        android:layout_alignParentTop="true"
        android:layout_centerHorizontal="true"
        android:textSize="30dp" />

    <ImageButton
        android:layout_width="wrap_content"
        android:layout_height="wrap_content"
        android:id="@+id/imageButton"
        android:src="@drawable/ic_launcher"
        android:layout_centerHorizontal="true"
        android:layout_below="@+id/textView1" />

    <Button
        android:layout_width="wrap_content"
        android:layout_height="wrap_content"
        android:text="Start Browser"
        android:id="@+id/button"
        android:layout_alignLeft="@+id/imageButton"
        android:layout_alignStart="@+id/imageButton"
        android:layout_below="@+id/imageButton"
        android:layout_alignEnd="@+id/imageButton" />

    <Button
        android:layout_width="wrap_content"
        android:layout_height="wrap_content"
        android:text="Start Phone"
        android:id="@+id/button2"
        android:layout_below="@+id/button"
        android:layout_alignLeft="@+id/button"
```

```xml
            android:layout_alignStart="@+id/button"
            android:layout_alignEnd="@+id/imageButton" />
</RelativeLayout>
```

修改工程的 java/com.example.intent1 目录下 java 文件 MainActivity.java，修改代码如下：

```java
package com.example.intent1;

import android.app.Activity;
import android.content.Intent;
import android.net.Uri;
import android.os.Bundle;
import android.view.View;
import android.widget.Button;

public class MainActivity extends Activity {
    Button b1,b2;
    @Override
    protected void onCreate(Bundle savedInstanceState) {
        super.onCreate(savedInstanceState);
        setContentView(R.layout.activity_main);
        b1=(Button)findViewById(R.id.button);
        b1.setOnClickListener(new View.OnClickListener() {

            @Override
            public void onClick(View v) {
                Intent i = new Intent(android.content.Intent.ACTION_VIEW,
                        Uri.parse("http://www.baidu.com"));
                startActivity(i);
            }
        });

        b2=(Button)findViewById(R.id.button2);
        b2.setOnClickListener(new View.OnClickListener() {
            @Override
            public void onClick(View v) {
                Intent i = new Intent(android.content.Intent.ACTION_VIEW,
                        Uri.parse("tel:18611111186"));
                startActivity(i);
            }
        });
    }
}
```

在 Genymotion 中运行程序的效果如图 4.4 所示。
然后单击"START BROWSER"按钮，运行效果如图 4.5 所示。
然后单击"START PHONE"按钮，运行效果如图 4.6 所示。

第 4 章 Android 应用核心 Intent 和 Filters

图 4.4 Intent（意图）实例 1 运行效果 1

图 4.5 Intent（意图）实例 1 运行效果 2

图 4.6 Intent（意图）实例 1 运行效果 3

4.3 Intent Filters（意图过滤器）

Activity、Service 和 BroadcastReceiver 能定义多个 Intent 过滤器来通知系统它们可以处理哪些隐式 Intent。每个过滤器描述组件的一种能力以及该组件可以接收的一组 Intent。实际上，过滤器接收需要类型的 Intent，拒绝不需要类型的 Intent 仅限于隐式 Intent。对于显式 Intent，无论内容如何，总可以发送给目标，过滤器不干预。

过滤器使用<intent-filter.../>标签在配置文件 AndroidManifest.xml 中进行注册，它是<activity.../>标签的子元素，<activity.../>标签用于为应用程序配置 Activity，<activity.../>的<intent-filter.../>标签则用于配置该 Activity 所能"响应"的 Intent。

<intent-filter.../>标签里通常包含如下子标签。
- 0～N 个<action.../>子标签。
- 0～N 个<category.../>子标签。
- 0～1 个<data.../>子标签。

以下面这个<activity.../>的标签为例：

```xml
<activity android:name=".CustomActivity"
    android:label="@string/app_name">
    <intent-filter>
        <action android:name="android.intent.action.VIEW" />
        <action android:name="com.example.intentdemo.LAUNCH" />
        <category android:name="android.intent.category.DEFAULT" />
        <data android:scheme="http" />
    </intent-filter>
</activity>
```

如上所示，这个 Activity 过滤器是有两个<action…/>子标签，表明其他 Activity 调用该 Activity 时要么使用 android.intent.action.VIEW 动作，要么使用 com.example.intentdemo.LAUNCH 动作。

一个<category…/>子标签表示类别为 android.intent.category.DEFAULT。

一个<data…/>子标签表示需要 Activity 启动的数据类型为 http://。

那么，如果一个 Intent 可以通过多个 Activity 或 Service 的<intent-filter>标签，系统到底会调用哪一个 Activity 或 Service 组件来执行这个 Intent 呢？

Android 系统会通过如下的测试来决定调用哪一个组件。

- <intent-filter>标签可以列有多个<action…/>子标签，系统会选择其中最匹配的动作来调用 Activity。但<action…/>子标签不能缺省，即过滤器必须包含至少一个<action…/>子标签，否则会阻塞所有 Intent。
- <intent-filter>标签可以列 0 个、1 个或多个<category…/>子标签。0 个表示 Intent 总是可以通过该项测试。1 个或多个表示 Intent 类别必须和每一种类别相匹配才能通过测试。需要注意的是，Android 默认所有通过 startActivity()方法传递的隐式 Intent 包含一个默认类型 android.intent.category.DEFAULT（CATEGORY_DEFAULT 常量），因此，接收隐式 Intent 的 Activity 必须在过滤器中包含 android.intent.category.DEFAULT 相与之匹配。
- <intent-filter>标签可以列 0 个、1 个或多个< data …/>子标签。每个< data …/>子标签可以指定 URI 和数据类型（MIME 媒体类型）。URI 可以分成 scheme、host、port 和 path 几个独立的部分。Intent 对象的 URI 和数据类型两部分都要和过滤器的 URI 和数据类型相互匹配才能通过测试，匹配规则如表 4.9 所示。

表 4.9 <data …/>子标签测试规则说明

Intent 对象		过滤器		通过条件
URI	数据类型	URI	数据类型	
未指定	未指定	未指定	未指定	无条件通过
指定	未指定	指定	未指定	两个 URI 匹配
未指定	指定	未指定	指定	两个数据类型匹配
指定	指定	指定	指定	URI 和数据类型匹配

【例 4.2】 Intent 使用过滤器实例

这个案例将简单修改上一个 Intent（意图）实例，我们将看到一个 Intent 如何调用两个 Activity 让用户做出选择的例子、一个用过滤器测试 Intent 调用 Activity 的例子和一个 Intent 没有适合 Activity 调用出错的例子。Intent（意图）2 实例步骤如表 4.10 所示。

表 4.10 Intent（意图）2 实例步骤

步骤	描述
1	打开 Android Studio 创建一个 Android 应用，Application name（应用名）取名为 Intent2，Company Domain（公司域）取名为 Example.com
2	Minimum SDK 选择 API 18: Android 4.3（Jelly Bean）
3	选择 Empty Activity
4	不需要修改 Activity Name，使用默认值，单击"Finish"按钮（完成）
5	将工程 res/mipmap 目录下的 ic_launcher.png 图片文件复制、粘贴到 res/drawable 目录下

步骤	描述
6	在工程中找到 ras/layout 目录中的 activity_main.xml 文件，在其中添加一个 TextView（文本框）控件、一个 ImageButton（图片按钮）和三个 Button（普通按钮）控件（Start browsing with view action、Start browsing with launch action 和 Exceptional condition），按图 4.7 所示的运行效果位置进行摆放
7	修改 java/com.example.intent2 目录下 java 文件 MainActivity.java 代码，为 Start browsing with view action、Start browsing with launch action 和 Exceptional condition 三个按钮添加单击事件监听器
8	右键单击 java/com.example.intent2 目录，选择 "new" → "Activity" → "Empty Activity"，新建一个 Activity 文件命名为：CustomActivity.java，按下面的代码修改文件
9	在工程中找到 ras/layout 目录中随 CustomActivity 新生成的 activity_custom.xml 文件，在其中添加一个 TextView（文本框）控件，将用于显示 Intent 传来的值
10	按下面的代码修改 manifests 目录下的配置文件 AndroidManifest.xml
11	启动 Genymotion 模拟器，然后在 Android 工程中做如下代码修改

修改新建工程 res/layout 目录下的布局文件 activity_main.xml，修改代码如下：

```xml
<?xml version="1.0" encoding="utf-8"?>
<RelativeLayout xmlns:android="http://schemas.android.com/
        apk/res/android"
    xmlns:tools="http://schemas.android.com/tools" android:layout_width=
        "match_parent"
    android:layout_height="match_parent" android:paddingLeft="@dimen/
        activity_horizontal_margin"
    android:paddingRight="@dimen/activity_horizontal_margin"
    android:paddingTop="@dimen/activity_vertical_margin"
    android:paddingBottom="@dimen/activity_vertical_margin" tools:context=
        ".MainActivity">

    <TextView
        android:id="@+id/textView1"
        android:layout_width="wrap_content"
        android:layout_height="wrap_content"
        android:text="Intent Example"
        android:layout_alignParentTop="true"
        android:layout_centerHorizontal="true"
        android:textSize="30dp" />

    <ImageButton
        android:layout_width="wrap_content"
        android:layout_height="wrap_content"
        android:id="@+id/imageButton"
        android:src="@drawable/ic_launcher"
        android:layout_centerHorizontal="true"
        android:layout_below="@+id/textView1" />

    <Button
        android:layout_width="wrap_content"
        android:layout_height="wrap_content"
        android:text="Start browsing with view action"
```

```xml
        android:id="@+id/button"
        android:layout_alignLeft="@+id/imageButton"
        android:layout_alignStart="@+id/imageButton"
        android:layout_below="@+id/imageButton"
        android:layout_alignEnd="@+id/imageButton" />

    <Button
        android:layout_width="wrap_content"
        android:layout_height="wrap_content"
        android:text="Start browsing with launch action"
        android:id="@+id/button2"
        android:layout_below="@+id/button"
        android:layout_alignLeft="@+id/button"
        android:layout_alignStart="@+id/button"
        android:layout_alignEnd="@+id/button" />

    <Button
        android:layout_width="wrap_content"
        android:layout_height="wrap_content"
        android:text="Exceptional condition"
        android:id="@+id/button3"
        android:layout_below="@+id/button2"
        android:layout_alignLeft="@+id/button2"
        android:layout_alignStart="@+id/button2"
        android:layout_alignEnd="@+id/button2" />
</RelativeLayout>
```

修改工程的 java/com.example.intent2 目录下 java 文件 MainActivity.java，修改代码如下：

```java
package com.example.intent2;

import android.app.Activity;
import android.content.Intent;
import android.net.Uri;
import android.os.Bundle;
import android.view.View;
import android.widget.Button;

public class MainActivity extends Activity {
    Button b1,b2,b3;
    @Override
    protected void onCreate(Bundle savedInstanceState) {
        super.onCreate(savedInstanceState);
        setContentView(R.layout.activity_main);
        b1=(Button)findViewById(R.id.button);

        b1.setOnClickListener(new View.OnClickListener() {
```

```java
            @Override
            public void onClick(View v) {
                Intent i = new Intent(android.content.Intent.ACTION_VIEW,Uri.parse
                        ("http://www.baidu.com"));
                startActivity(i);
            }
        });

        b2=(Button)findViewById(R.id.button2);
        b2.setOnClickListener(new View.OnClickListener() {

            @Override
            public void onClick(View v) {
                Intent i = new Intent("com.example.intent2.LAUNCH", Uri.parse
                        ("http://www.baidu.com"));
                startActivity(i);
            }
        });

        b3=(Button)findViewById(R.id.button3);
        b3.setOnClickListener(new View.OnClickListener() {

            @Override
            public void onClick(View v) {
                Intent i = new Intent("com.example.intent2.LAUNCH",
                        Uri.parse("https://www.baidu.com"));
                startActivity(i);
            }
        });
    }
}
```

修改工程 res/layout 目录下的布局文件 activity_custom.xml，修改代码如下：

```xml
<?xml version="1.0" encoding="utf-8"?>
<RelativeLayout xmlns:android="http://schemas.android.com/
                apk/res/android"
    xmlns:tools="http://schemas.android.com/tools" android:layout_width=
                "match_parent"
    android:layout_height="match_parent" android:paddingLeft=
                "@dimen/activity_horizontal_margin"
    android:paddingRight="@dimen/activity_horizontal_margin"
    android:paddingTop="@dimen/activity_vertical_margin"
    android:paddingBottom="@dimen/activity_vertical_margin"
    tools:context="com.example.intent2.CustomActivity">
    <TextView android:id="@+id/show_data"
        android:layout_width="fill_parent"
        android:layout_height="400dp"/>
</RelativeLayout>
```

修改工程的 java/com.example.intent2 目录下 java 文件 CustomActivity.java，修改代码如下：

```java
package com.example.intent2;

import android.net.Uri;
import android.os.Bundle;
import android.support.v7.app.AppCompatActivity;
import android.widget.TextView;

public class CustomActivity extends AppCompatActivity {

    @Override
    protected void onCreate(Bundle savedInstanceState) {
        super.onCreate(savedInstanceState);
        setContentView(R.layout.activity_custom);
        TextView label = (TextView) findViewById(R.id.show_data);
        Uri url = getIntent().getData();
        label.setText(url.toString());
    }
}
```

修改工程的 manifests 目录下配置文件 AndroidManifest.xml，修改代码如下：

```xml
<?xml version="1.0" encoding="utf-8"?>
<manifest xmlns:android="http://schemas.android.com/apk/res/android"
    package="com.example.intent2" >

    <application
        android:allowBackup="true"
        android:icon="@mipmap/ic_launcher"
        android:label="@string/app_name"
        android:supportsRtl="true"
        android:theme="@style/AppTheme" >
        <activity android:name=".MainActivity" >
            <intent-filter>
                <action android:name="android.intent.action.MAIN" />

                <category android:name="android.intent.category.LAUNCHER" />
            </intent-filter>
        </activity>

        <activity android:name=".CustomActivity" >
            <intent-filter>
                <action android:name="android.intent.action.VIEW" />
                <action android:name="com.example.intent2.LAUNCH" />
                <category android:name="android.intent.category.DEFAULT" />
                <data android:scheme="http" />
            </intent-filter>
        </activity>
    </application>

</manifest>
```

在 Genymotion 中运行程序的效果如图 4.7 所示。

现在，首先单击"Start browsing with view action"按钮。由于我们在 CustomActivity 的过滤器中定义了如下两个动作标签，<action android:name="android.intent.action.VIEW" />标签表示系统会调用网页浏览器来响应 Intent，<action android:name="com.example.intent2.LAUNCH" />标签则表示用 CustomActivity 来响应 Intent，两者都是可行的。所以将会打开如图 4.8 所示的选择对话框，让用户选择一个组件来响应该 Intent。

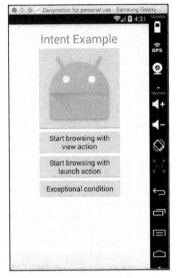

图 4.7　Intent（意图）实例 2 运行效果

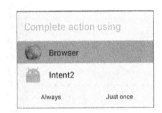

图 4.8　用户选择 Intent 打开组件对话框

如果用户选择了"Browser"，那么 Android 将启动网页浏览器打开网址"http://www.baidu.com"，如图 4.9 所示。

如果用户选择了"Intent2"，那么 Android 将调用 CustomActivity 组件并把 Intent 值"http://www.baidu.com"传递给其中的文本框，如图 4.10 所示。

图 4.9　用户选择系统组件响应 Intent 请求

图 4.10　用户选择 CustomActivity 组件响应 Intent 请求

单击"模拟器 Back"按钮，退回实例主界面。单击"Start browsing with launch action"按钮，这时 Android 启用过滤器对组件进行测试，请求 Intent 与 CustomActivity 组件过滤器相匹配，所以系统直接调用 CustomActivity 组件并把 Intent 值"http://www.baidu.com"传递给其中的文本框，运行效果如图 4.11 所示。

再单击"模拟器 Back"按钮，退回实例主界面。单击"Exceptional condition"按钮，由于我们把 Intent 中的请求数据 http 修改成了 https，Android 系统不能找到与之匹配的组件，所以系统将提示错误，应用程序将被退出，如图 4.12 所示。

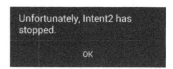

图 4.11　请求 Intent 与 CustomActivity 组件过滤器相匹配　　图 4.12　系统找不到与 Intent 请求相匹配的组件而报错

第 5 章　Android 事件处理

不管是桌面应用还是手机应用程序，面对最多的就是用户，经常需要处理的就是用户动作，也就是需要响应用户动作，这种响应用户动作的机制就是事件处理。

5.1　Android 事件处理概述

下面介绍 Android 事件管理的三个基本概念。
- 事件监听器：事件监听器就是包含一个回调方法的 View（视图）类接口，它负责监听 UI 界面上的 View（视图）类（控件）发生的各种事件，并对事件做出相应的响应。
- 事件监听器注册：事件监听器注册就是事件监听器与一个事件处理程序相绑定，当事件发生时，监听器侦测到事件后交由与其绑定（注册）的事件处理程序做响应处理。
- 事件处理程序：当事件发生时，事件监听器将调用与其绑定（注册）的事件处理程序以处理事件。事件处理程序中包含处理事件的实例方法。

常见的事件处理程序与事件监听器如表 5.1 所示。

表 5.1　常见的事件处理程序与事件监听器

事件处理程序	事件监听器
onClick()	OnClickListener()： 当用户单击、触摸或把焦点放到按钮、文本或图片等控件上时，它将调用 onClick() 事件处理程序来响应该事件
onLongClick()	OnLongClickListener()： 当用户单击、触摸或把焦点放到按钮、文本或图片等控件上的时间超过 1s 时，它将调用 onLongClick() 事件处理程序来响应该事件
onFocusChange()	OnFocusChangeListener()： 当控件失去焦点时，它将调用 onFocusChange() 事件处理程序来响应该事件
onTouch()	OnTouchListener()： 当用户按下硬件按键、松开硬件按键或在屏幕上做任何手势时，它将调用 onTouch() 事件处理程序来响应该事件
onMenuItemClick()	OnMenuItemClickListener()： 当用户选中一个菜单项时，它将调用 onMenuItemClick() 事件处理程序来响应该事件

注意：在 Android 中还有很多其他的监听器，如 OnHoverListener、OnDragListener 等，但这里并没有列出来。当开发一个复杂应用程序时，请查看 Android 官方 API 文档。

5.2　注册事件监听器的方法

事件监听器注册就是事件监听器与一个事件处理程序相绑定。注册方法有很多种，但最常用的方法有以下三种。
- 使用匿名内部类。
- Activity 类执行 Listener（监听器）接口。

- 在布局文件 activity_main.xml 中直接指定事件处理程序(方法)。

下面举例说明以上三种情景。

【例 5.1】 基于匿名内部类的事件监听器注册

这个例子将通过几个简单的步骤展示如何使用 OnClickListener 类注册并捕获单击事件。基于匿名内部类的事件监听器注册案例实例步骤如表 5.2 所示。

表 5.2 基于匿名内部类的事件监听器注册案例实例步骤

步骤	描述
1	打开 Android Studio 创建一个 Android 应用,Application name(应用名)取名为 EventDemo1,Company Domain(公司域)取名为 Example.com
2	Minimum SDK 选择 API 18: Android 4.3 (Jelly Bean)
3	选择 Empty Activity
4	不需要修改 Activity Name,使用默认值,单击"Finish"按钮(完成)
5	将工程 res/mipmap 目录下的 ic_launcher.png 图片文件复制、粘贴到 res/drawable 目录下
6	在工程中找到 ras/layout 目录中的 activity_main.xml 文件,在其中添加一个 TextView(文本框)控件、一个 ImageButton(图片按钮)和两个 Button(普通按钮)控件(Small font 和 Large Font),按图 5.1 所示的运行效果位置进行摆放
7	修改 java/com.example.eventdemo1 目录下 java 文件 MainActivity.java 代码,为 Small font 和 Large Font 两个按钮添加单击事件监听器

修改新建工程的 res/layout 目录下的布局文件 activity_main.xml,修改代码如下:

```xml
<?xml version="1.0" encoding="utf-8"?>
<RelativeLayout xmlns:android="http://schemas.android.com/
        apk/res/android"
    xmlns:tools="http://schemas.android.com/tools" android:layout_width=
        "match_parent"
    android:layout_height="match_parent" android:paddingLeft="@dimen/
        activity_horizontal_margin"
    android:paddingRight="@dimen/activity_horizontal_margin"
    android:paddingTop="@dimen/activity_vertical_margin"
    android:paddingBottom="@dimen/activity_vertical_margin"
        tools:context=".MainActivity">

    <TextView
        android:id="@+id/textView1"
        android:layout_width="wrap_content"
        android:layout_height="wrap_content"
        android:text="Event Handling"
        android:layout_alignParentTop="true"
        android:layout_centerHorizontal="true"
        android:textSize="30dp"/>

    <ImageButton
        android:layout_width="wrap_content"
        android:layout_height="wrap_content"
        android:id="@+id/imageButton"
        android:src="@drawable/ic_launcher"
```

```xml
        android:layout_centerVertical="true"
        android:layout_centerHorizontal="true" />

    <Button
        android:layout_width="wrap_content"
        android:layout_height="wrap_content"
        android:text="Small font"
        android:id="@+id/button"
        android:layout_below="@+id/imageButton"
        android:layout_alignEnd="@+id/textView1"
        android:layout_alignStart="@+id/textView1" />

    <Button
        android:layout_width="wrap_content"
        android:layout_height="wrap_content"
        android:text="Large Font"
        android:id="@+id/button2"
        android:layout_below="@+id/button"
        android:layout_alignRight="@+id/button"
        android:layout_alignEnd="@+id/button"
        android:layout_alignStart="@+id/button" />

    <TextView
        android:layout_width="wrap_content"
        android:layout_height="wrap_content"
        android:text="Hello World!"
        android:id="@+id/textView"
        android:layout_below="@+id/button2"
        android:layout_centerHorizontal="true"
        android:textSize="25dp" />
</RelativeLayout>
```

修改工程的 java/com.example.eventdemo1 目录下 java 文件 MainActivity.java,修改代码如下:

```java
package com.example.eventdemo1;

import android.os.Bundle;
import android.support.v7.app.AppCompatActivity;
import android.view.View;
import android.widget.Button;
import android.widget.TextView;

public class MainActivity extends AppCompatActivity {

    @Override
    protected void onCreate(Bundle savedInstanceState) {
        super.onCreate(savedInstanceState);
        setContentView(R.layout.activity_main);
```

```
//--- find both the buttons---
Button sButton = (Button) findViewById(R.id.button);
Button lButton = (Button) findViewById(R.id.button2);

// -- register click event with first button ---
sButton.setOnClickListener(new View.OnClickListener() {
    public void onClick(View v) {
        // --- find the text view --
        TextView txtView = (TextView) findViewById(R.id.textView);
        // -- change text size --
        txtView.setTextSize(14);
    }
});

// -- register click event with second button ---
lButton.setOnClickListener(new View.OnClickListener() {
    public void onClick(View v) {
        // --- find the text view --
        TextView txtView = (TextView) findViewById(R.id.textView);
        // -- change text size --
        txtView.setTextSize(24);
    }
});
    }
}
```

在 Genymotion 中运行程序的效果如图 5.1 所示。

图 5.1 基于匿名内部类的事件监听器注册实例运行效果

【例 5.2】 Activity 类执行 Listener 接口注册监听器

这个例子将通过几个简单的步骤展示如何使用 Activity 类执行 Listener 接口注册监听器并捕获单击事件。Activity 类执行 Listener 接口注册监听器案例实例步骤如表 5.3 所示。

表 5.3 Activity 类执行 Listener 接口注册监听器案例实例步骤

步骤	描述
1	打开 Android Studio 创建一个 Android 应用，Application name（应用名）取名为 EventDemo2，Company Domain（公司域）取名为 Example.com
2	Minimum SDK 选择 API 18: Android 4.3（Jelly Bean）
3	选择 Empty Activity
4	不需要修改 Activity Name，使用默认值，单击"Finish"按钮（完成）
5	将工程 res/mipmap 目录下的 ic_launcher.png 图片文件复制、粘贴到 res/drawable 目录下
6	在工程中找到 ras/layout 目录中的 activity_main.xml 文件，在其中添加一个 TextView（文本框）控件、一个 ImageButton（图片按钮）和两个 Button（普通按钮）控件（Small font 和 Large Font），按图 5.2 所示的运行效果位置进行摆放
7	按以下代码修改 java/com.example.eventdemo2 目录下 java 文件 MainActivity.java

修改新建工程的 res/layout 目录下的布局文件 activity_main.xml，修改代码如下：

```xml
<?xml version="1.0" encoding="utf-8"?>
<RelativeLayout xmlns:android="http://schemas.android.com/
    apk/res/android"
    xmlns:tools="http://schemas.android.com/tools" android:layout_width=
        "match_parent"
    android:layout_height="match_parent" android:paddingLeft=
        "@dimen/activity_horizontal_margin"
    android:paddingRight="@dimen/activity_horizontal_margin"
    android:paddingTop="@dimen/activity_vertical_margin"
    android:paddingBottom="@dimen/activity_vertical_margin"
        tools:context=".MainActivity">

    <TextView
        android:id="@+id/textView1"
        android:layout_width="wrap_content"
        android:layout_height="wrap_content"
        android:text="Event Handling"
        android:layout_alignParentTop="true"
        android:layout_centerHorizontal="true"
        android:textSize="30dp"/>

    <ImageButton
        android:layout_width="wrap_content"
        android:layout_height="wrap_content"
        android:id="@+id/imageButton"
        android:src="@drawable/ic_launcher"
        android:layout_centerVertical="true"
        android:layout_centerHorizontal="true" />
```

```xml
    <Button
        android:layout_width="wrap_content"
        android:layout_height="wrap_content"
        android:text="Small font"
        android:id="@+id/button"
        android:layout_below="@+id/imageButton"
        android:layout_alignEnd="@+id/imageButton"
        android:layout_alignStart="@+id/textView1" />

    <Button
        android:layout_width="wrap_content"
        android:layout_height="wrap_content"
        android:text="Large Font"
        android:id="@+id/button2"
        android:layout_below="@+id/button"
        android:layout_alignRight="@+id/button"
        android:layout_alignEnd="@+id/button"
        android:layout_alignStart="@+id/button" />

    <TextView
        android:layout_width="wrap_content"
        android:layout_height="wrap_content"
        android:text="Hello World!"
        android:id="@+id/textView"
        android:layout_below="@+id/button2"
        android:layout_centerHorizontal="true"
        android:textSize="25dp" />
</RelativeLayout>
```

修改工程的 java/com.example.eventdemo2 目录下 java 文件 MainActivity.java，修改代码如下：

```java
package com.example.eventdemo2;

import android.app.Activity;
import android.os.Bundle;
import android.view.View;
import android.widget.Button;
import android.widget.TextView;

public class MainActivity extends Activity implements View.OnClickListener {

    @Override
    protected void onCreate(Bundle savedInstanceState) {
        super.onCreate(savedInstanceState);
        setContentView(R.layout.activity_main);
        //--- find both the buttons---
        Button sButton = (Button) findViewById(R.id.button);
        Button lButton = (Button) findViewById(R.id.button2);
```

```
        //-- register click event with first button ---
        sButton.setOnClickListener(this);
        //-- register click event with second button ---
        lButton.setOnClickListener(this);
    }

    //--- Implement the OnClickListener callback
    public void onClick(View v) {
        if (v.getId() == R.id.button) {
            //--- find the text view --
            TextView txtView = (TextView) findViewById(R.id.textView);
            //-- change text size --
            txtView.setTextSize(14);
            return;
        }
        if (v.getId() == R.id.button2) {
            //--- find the text view --
            TextView txtView = (TextView) findViewById(R.id.textView);
            //-- change text size --
            txtView.setTextSize(24);
            return;
        }
    }
}
```

在 Genymotion 中运行程序的效果如图 5.2 所示。

图 5.2　Activity 类执行 Listener 接口注册监听器实例运行效果

【例 5.3】 使用布局文件 activity_main.xml 注册监听器

这个例子将通过几个简单的步骤展示如何使用布局文件 activity_main.xml 注册监听器并捕获单击事件。布局文件 activity_main.xml 注册监听器案例实例步骤如表 5.4 所示。

表 5.4 布局文件 activity_main.xml 注册监听器案例实例步骤

步骤	描述
1	打开 Android Studio 创建一个 Android 应用，Application name（应用名）取名为 EventDemo3，Company Domain（公司域）取名为 Example.com
2	Minimum SDK 选择 API 18: Android 4.3（Jelly Bean）
3	选择 Empty Activity
4	不需要修改 Activity Name，使用默认值，单击"Finish"按钮（完成）
5	将工程 res/mipmap 目录下的 ic_launcher.png 图片文件复制、粘贴到 res/drawable 目录下
6	在工程中找到 ras/layout 目录中的 activity_main.xml 文件，在其中添加一个 TextView（文本框）控件、一个 ImageButton（图片按钮）和两个 Button（普通按钮）控件（Small font 和 Large Font），按图 5.3 所示的运行效果位置进行摆放，并按以下代码修改
7	按以下代码修改 java/com.example.eventdemo3 目录下 java 文件 MainActivity.java

修改新建工程的 res/layout 目录下的布局文件 activity_main.xml，修改代码如下：

```xml
<?xml version="1.0" encoding="utf-8"?>
<RelativeLayout xmlns:android="http://schemas.android.com/
     apk/res/android"
    xmlns:tools="http://schemas.android.com/tools" android:layout_width=
        "match_parent"
    android:layout_height="match_parent" android:paddingLeft="@dimen/
        activity_horizontal_margin"
    android:paddingRight="@dimen/activity_horizontal_margin"
    android:paddingTop="@dimen/activity_vertical_margin"
    android:paddingBottom="@dimen/activity_vertical_margin" tools:context=
        ".MainActivity">

    <TextView
        android:id="@+id/textView1"
        android:layout_width="wrap_content"
        android:layout_height="wrap_content"
        android:text="Event Handling"
        android:layout_alignParentTop="true"
        android:layout_centerHorizontal="true"
        android:textSize="30dp"/>

    <ImageButton
        android:layout_width="wrap_content"
        android:layout_height="wrap_content"
        android:id="@+id/imageButton"
        android:src="@drawable/ic_launcher"
        android:layout_centerVertical="true"
        android:layout_centerHorizontal="true" />
```

```xml
    <Button
        android:layout_width="wrap_content"
        android:layout_height="wrap_content"
        android:text="Small font"
        android:id="@+id/button"
        android:onClick="doSmall"
        android:layout_below="@+id/imageButton"
        android:layout_alignEnd="@+id/imageButton"
        android:layout_alignStart="@+id/textView1" />

    <Button
        android:layout_width="wrap_content"
        android:layout_height="wrap_content"
        android:text="Large Font"
        android:id="@+id/button2"
        android:onClick="doLarge"
        android:layout_below="@+id/button"
        android:layout_alignRight="@+id/button"
        android:layout_alignEnd="@+id/button"
        android:layout_alignStart="@+id/button" />

    <TextView
        android:layout_width="wrap_content"
        android:layout_height="wrap_content"
        android:text="Hello World!"
        android:id="@+id/textView"
        android:layout_below="@+id/button2"
        android:layout_centerHorizontal="true"
        android:textSize="25dp" />
</RelativeLayout>
```

修改工程的 java/com.example.eventdemo3 目录下 java 文件 MainActivity.java，修改代码如下：

```java
package com.example.eventdemo3;

import android.os.Bundle;
import android.support.v7.app.AppCompatActivity;
import android.view.View;
import android.widget.TextView;

public class MainActivity extends AppCompatActivity {

    @Override
    protected void onCreate(Bundle savedInstanceState) {
        super.onCreate(savedInstanceState);
        setContentView(R.layout.activity_main);
    }
    //--- Implement the event handler for the first button.
```

```
            public void doSmall(View v)  {
                //--- find the text view --
                TextView txtView = (TextView) findViewById(R.id.textView);
                //-- change text size --
                txtView.setTextSize(14);
                return;
            }
            //--- Implement the event handler for the second button.
            public void doLarge(View v)  {
                //--- find the text view --
                TextView txtView = (TextView) findViewById(R.id.textView);
                //-- change text size --
                txtView.setTextSize(24);
                return;
            }
        }
```

在 Genymotion 中运行程序的效果如图 5.3 所示。

图 5.3　布局文件 activity_main.xml 注册监听器实例运行效果

第 6 章 Android 服务

Service(服务)是能够在后台执行长时间运行操作且不提供用户界面的应用程序组件。例如，Service 能在后台播放音乐、处理网络事务或执行文件输入/输出等操作。Service 也是在 Android 四大组件中与 Activity 最相似的组件，它们都代表可执行的程序，Service 与 Activity 的区别是，Service 一直在后台运行，没有用户界面，所以绝对不会到前台来。

6.1 Service 的分类

Service 按其状态可以分为两种类型，如表 6.1 所示。

表 6.1 Service 的分类

状态	描述
Started(启动)	当应用程序组件(如 Activity)通过调用 startService()方法启动服务时，服务处于 started 状态。一旦启动，Service 能在后台无限期地运行，即使启动它的组件已经被销毁。例如，Service 通过网络下载或者上传文件。如果操作完成，Service 需要停止运行
Bound(绑定)	当应用程序组件通过调用 bindService()方法绑定到 Service 时，服务处于 bound 状态。绑定 Service 提供客户端/服务器接口，以允许组件与服务交互、发送请求、获得结果，甚至使用进程间通信(IPC)跨进程完成这些操作。仅当其他应用程序与 Service 绑定时，绑定服务才运行。多个组件可以一次绑定到一个 Service 上，当它们都解绑时，Service 被销毁

6.2 Service 的生命周期

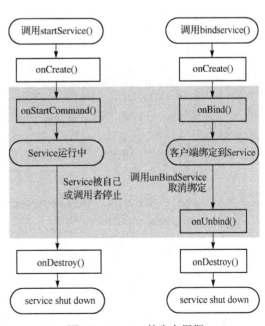

Service 与 Activity 一样具有自己的生命周期。通过生命周期中的回调方法，用户可以监视 Service 的状态变化。随着应用程序启动 Service 的方式不同，Service 的生命周期也略有差异。

如果应用程序通过 startService()方法启动 Service，Service 的生命周期如图 6.1 左部分所示。

如果应用程序通过 bindService()方法启动 Service，Service 的生命周期如图 6.1 右部分所示。

图 6.1 Service 的生命周期

6.3 Service 生命周期中的回调方法

为了创建 Service，需要创建一个 Service 类的子类。在实现类中，需要重写一些处理服务生命周期的回调方法，并根据需要提供组件绑定到服务的机制。需要重写的重要回调方法如表 6.2 所示。

表 6.2 Service 生命周期中的回调方法

回调方法	描述
onStartCommand()	当其他组件(如 Activity)调用 startService()方法启动该 Service 时，系统都会回调该方法。如果任务完成需要停止服务，可以通过 stopSelf()或 stopService()方法实现
onBind()	当其他组件(如 Activity)调用 bindService()方法想与 Service 绑定时，系统调用该方法。在该方法的实现中，开发人员必须通过返回 IBinder 对象来提供其他组件与 Service 通信的接口。该方法必须实现，但是如果不想绑定，则需返回 null
onUnbind()	当该 Service 上绑定的所有其他组件都断开连接时将会回调该方法
onRebind()	当该 Service 和旧组件之间的所有的绑定在 onUnbind()里面全都结束之后，如果一个新的组件要绑定到该 Service，就会启动该方法
onCreate()	当该 Service 第一次被创建后将立即回调该方法。如果 Service 已经运行，该方法不被调用
onDestroy()	在该 Service 被关闭之前将会回调该方法

下面的类定义了一个 Service 组件，重写了 Service 组件的 onCreate()、onStartCommand()、onBind()、onUnbind()、onRebind()和 onDestroy()方法，重写这些方法时只是简单地输出了一条状态信息的字符串。

```java
package com.example.helloservice;

import android.app.Service;
import android.content.Intent;
import android.os.IBinder;

public class HelloService extends Service {
    /** indicates how to behave if the service is killed */
    int mStartMode;

    /** interface for clients that bind */
    IBinder mBinder;

    /** indicates whether onRebind should be used */
    boolean mAllowRebind;

    /** Called when the service is being created. */
    @Override
    public void onCreate() {

    }

    /** The service is starting, due to a call to startService() */
```

```java
@Override
public int onStartCommand(Intent intent, int flags, int startId) {
    return mStartMode;
}

/** A client is binding to the service with bindService() */
@Override
public IBinder onBind(Intent intent) {
    return mBinder;
}

/** Called when all clients have unbound with unbindService() */
@Override
public boolean onUnbind(Intent intent) {
    return mAllowRebind;
}

/** Called when a client is binding to the service with bindService()*/
@Override
public void onRebind(Intent intent) {

}

/** Called when The service is no longer used and is being destroyed */
@Override
public void onDestroy() {

}
}
```

【例 6.1】 Service 实例

这个例子将通过几个简单的步骤展示如何创建一个 Android 服务,如表 6.3 所示。

表 6.3 创建 Service 案例实例步骤

步骤	描述
1	打开 Android Studio 创建一个 Android 应用,Application name(应用名)取名为 HelloServiceDemo,Company Domain(公司域)取名为 Example.com
2	Minimum SDK 选择 API 18: Android 4.3 (Jelly Bean)
3	选择 Empty Activity
4	不需要修改 Activity Name,使用默认值,单击 "Finish" 按钮(完成)
5	右键单击工程的 com.example.helloservicedemo 目录,选择 "New" → "Service" → "Service",新建一个服务类取名为 MyService.java。该文件将实现 Android 服务的相关方法
6	在工程中找到 ras/layout 目录中的 activity_main.xml 文件,在其中添加两个 Button(普通按钮)控件(Start Service 和 Stop Service),按图 6.2 所示的运行效果位置进行摆放
7	按照如下代码修改 java/com.example.helloservicedemo 目录下 java 文件 MainActivity.java 代码,为其添加 startService() 和 stopService() 方法
8	使用 <service.../> 标签在 AndroidManifest.xml 文件中注册服务
9	启动 Genymotion 模拟器,然后在 Android 工程中做如下代码修改

修改工程的 res/layout 目录下的布局文件 activity_main.xml，修改代码如下：

```xml
<?xml version="1.0" encoding="utf-8"?>
<LinearLayout xmlns:android="http://schemas.android.com/apk/res/android"
    android:layout_width="fill_parent"
    android:layout_height="fill_parent"
    android:orientation="vertical" >

    <Button android:id="@+id/btnStartService"
        android:layout_width="fill_parent"
        android:layout_height="wrap_content"
        android:text="Start Service"
        android:onClick="startService"/>

    <Button android:id="@+id/btnStopService"
        android:layout_width="fill_parent"
        android:layout_height="wrap_content"
        android:text="Stop Service"
        android:onClick="stopService" />

</LinearLayout>
```

修改工程 java/com.example.helloservicedemo 目录下的服务类 java 文件 MyService.java，修改代码如下：

```java
package com.example.helloservicedemo;

import android.app.Service;
import android.content.Intent;
import android.os.IBinder;
import android.widget.Toast;

public class MyService extends Service {
    @Override
    public IBinder onBind(Intent arg0) {
        return null;
    }

    @Override
    public int onStartCommand(Intent intent, int flags, int startId) {
        //Let it continue running until it is stopped.
        Toast.makeText(this, "Service Started", Toast.LENGTH_LONG).show();
        return START_STICKY;
    }
    @Override
    public void onDestroy() {
        super.onDestroy();
        Toast.makeText(this, "Service Destroyed", Toast.LENGTH_LONG).show();
    }
}
```

修改工程 java/com.example.helloservicedemo 目录下的 java 文件 MainActivity.java，修改代码如下：

```java
package com.example.helloservicedemo;

import android.app.Activity;
import android.content.Intent;
import android.os.Bundle;
import android.view.View;

public class MainActivity extends Activity {

    @Override
    protected void onCreate(Bundle savedInstanceState) {
        super.onCreate(savedInstanceState);
        setContentView(R.layout.activity_main);
    }
    //Method to start the service
    public void startService(View view) {
        startService(new Intent(getBaseContext(), MyService.class));
    }

    //Method to stop the service
    public void stopService(View view) {
        stopService(new Intent(getBaseContext(), MyService.class));
    }
}
```

修改工程 manifests 目录下的配置文件 AndroidManifest.xml，修改代码如下：

```xml
<?xml version="1.0" encoding="utf-8"?>
<manifest xmlns:android="http://schemas.android.com/apk/res/android"
    package="com.example.helloservicedemo" >

    <application
        android:allowBackup="true"
        android:icon="@mipmap/ic_launcher"
        android:label="@string/app_name"
        android:supportsRtl="true"
        android:theme="@style/AppTheme" >
        <activity android:name=".MainActivity" >
            <intent-filter>
                <action android:name="android.intent.action.MAIN" />

                <category android:name="android.intent.category.LAUNCHER" />
            </intent-filter>
        </activity>

        <service
```

```
            android:name=".MyService"
            android:enabled="true"
            android:exported="true" >
        </service>
    </application>

</manifest>
```

在 Genymotion 中运行,单击"Start Service"按钮,效果如图 6.2 所示。
然后单击"Stop Service"按钮,效果如图 6.3 所示。

图 6.2　Service(服务)实例运行效果 1

图 6.3　Service(服务)实例运行效果 2

第 7 章　Android 广播接收器

Broadcast Receiver（广播接收器）也是 Android 系统四大组件之一，这种组件本质上就是一个全局的监听器，用于监听系统全局的广播信息。这些信息就是程序（包括用户开发的程序和系统内建的程序）所发出的 Broadcast Intent（广播意图）。比如，当从网上下载数据完成后，应用程序可以发出广播信息通知其他程序，数据已经下载完成并可以使用了。这时，Broadcast Receiver 可以监听到这个信息并采取对应的行动。

我们需要通过以下两步来创建 Broadcast Receiver，并使其响应与之相配的系统 Broadcast Intent：
- 创建 Broadcast Receiver。
- 注册 Broadcast Receiver。

也有可能在以上两步之间增加一个步骤，就是当执行用户自定义 Broadcast Intent 时，需要创建这个自定义 Intent，然后将其与 Broadcast Receiver 注册。

7.1　创建 Broadcast Receiver

可以通过创建一个 Broadcast Receiver 子类的方法来创建一个广播接收器。然后，再重写它的 onReceive() 方法，将 Intent 作为其参数来传递消息，例如以下代码：

```java
public class MyReceiver extends BroadcastReceiver {
  @Override
  public void onReceive(Context context, Intent intent) {
    Toast.makeText(context, "Intent Detected.", Toast.LENGTH_LONG).show();
  }
}
```

7.2　注册 Broadcast Receiver

一旦实现了 Broadcast Receiver，接下来就应该指定该 Broadcast Receiver 能匹配的 Intent，这就是注册 Broadcast Receiver，如图 7.1 所示。注册 Broadcast Receiver 又需要考虑以下两种情况：
- 接收用户自定义 Broadcast Intent 消息。
- 接收系统广播消息。

图 7.1　Broadcast Receiver

7.2.1 接收用户自定义 Broadcast Intent 消息

如果你的应用程序组件（如 Activity）生成一个自定义 Broadcast Intent，则可以通过 sendBroadcast() 方法将其发送出去，例如以下代码：

```java
Intent intent = new Intent();
intent.setAction("com.example.CUSTOM_INTENT");
sendBroadcast(intent);
```

在 AndroidManifest.xml 文件中对与之匹配的 Broadcast Receiver 进行注册，例如以下代码：

```xml
<application
    android:icon="@drawable/ic_launcher"
    android:label="@string/app_name"
    android:theme="@style/AppTheme" >
    <receiver android:name="MyReceiver">

        <intent-filter>
            <action android:name="com.example.CUSTOM_INTENT">
            </action>
        </intent-filter>

    </receiver>
</application>
```

每次 Broadcast 事件发生后，系统就会创建对应的 Broadcast Receiver 实例，并自动触发它的 onReceive() 方法，onReceive() 方法执行完后，Broadcast Receiver 实例就会被销毁。

如果 Broadcast Receiver 的 onReceive() 方法不能在 10s 内执行完成，Android 就会认为该程序无响应。所以，不要在 Broadcast Receiver 的 onReceive() 方法中执行一些耗时的操作，否则会弹出程序无响应对话框。如果确实需要根据 Broadcast 来完成一项比较耗时的操作，则可以考虑通过 Intent 启动一个 Service 来完成该操作。

7.2.2 接收系统广播消息

除了接收用户发送的广播之外，Broadcast Receiver 还有一个重要的用途：接收系统广播。如果应用程序需要在系统特定时刻执行某些操作，就可以通过监听系统广播来实现。注册监听系统广播的 Broadcast Receiver，需要在 AndroidManifest.xml 文件中对与系统广播 Action 常量匹配的 Broadcast Receiver 进行注册，例如以下代码：

```xml
<application
    android:icon="@drawable/ic_launcher"
    android:label="@string/app_name"
    android:theme="@style/AppTheme" >
    <receiver android:name="MyReceiver">

        <intent-filter>
            <action android:name="android.intent.action.BOOT_COMPLETED">
            </action>
```

```
            </intent-filter>

        </receiver>
    </application>
```

Android 的大量系统事件都会对外发送标准广播。Android 常见的广播 Action 常量如表 7.1 所示。

表 7.1　常见的广播 Action 常量

广播 Action 常量	描述
android.intent.action.BATTERY_CHANGED	电池电量改变的广播
android.intent.action.BATTERY_LOW	电池电量低的广播
android.intent.action.BATTERY_OKAY	表示电池电量充足，即从电池电量低变化到饱满时会发出广播
android.intent.action.BOOT_COMPLETED	当系统启动完成后，这个动作被广播一次（只有一次）
android.intent.action.SCREEN_ON	当屏幕被打开时的广播
android.intent.action.CALL	当用户根据指明的信息向某人拨打电话时的广播
android.intent.action.CALL_BUTTON	当用户单击"拨号"按钮时的广播
android.intent.action.DATE_CHANGED	当系统日期被改变时的广播
android.intent.action.REBOOT	当系统重启时的广播

注意：更多的 Action 常量请参考 Android 官方 API 文档中关于 Intent 的说明。

【例 7.1】　监听用户自定义 Broadcast Intent 消息实例

这个例子将通过几个简单的步骤展示如何创建一个 Android Broadcast Receiver 用于监听用户自定义 Broadcast Intent 消息，如表 7.2 所示。

表 7.2　监听用户自定义 Broadcast Intent 消息案例实例步骤

步骤	描述
1	打开 Android Studio 创建一个 Android 应用，Application name（应用名）取名为 BroadcastCustom，Company Domain（公司域）取名为 Example.com
2	Minimum SDK 选择 API 18: Android 4.3 (Jelly Bean)
3	选择 Empty Activity
4	不需要修改 Activity Name，使用默认值，单击"Finish"按钮（完成）
5	右键单击工程的 com.example.broadcastcustom 目录，选择"New"→"Other"→"Broadcast Receiver"，新建一个 Broadcast Receiver 类取名 MyReceiver.java。该文件将定义一个 Broadcast Receiver
6	将工程 res/mipmap 目录下的 ic_launcher.png 图片文件复制、粘贴到 res/drawable 目录下
7	在工程中找到 ras/layout 目录中的 activity_main.xml 文件，在其中添一个 textView（文本框控件）控件，一天 imageButton（图像按钮）和一个 Button（普通按钮）控件（Broadcast Intent），按图 7.2 所示的运行效果位置进行摆放
8	按照如下代码修改 java/com.example.broadcastcustom 目录下 java 文件 MainActivity.java 代码，为其添加自定义 Broadcast Intent 消息，并发送
9	使用<receiver.../>标签在 AndroidManifest.xml 文件中注册 Broadcast Receiver，并用<intent-filter.../>标签设置 Broadcast Receiver 的过滤器
10	启动 Genymotion 模拟器，然后在 Android 工程中做如下代码修改

修改工程 java/com.example.broadcastcustom 目录下的服务类 java 文件 MyReceiver.java，修改代码如下：

```
package com.example.broadcastcustom;
```

```java
import android.content.BroadcastReceiver;
import android.content.Context;
import android.content.Intent;
import android.widget.Toast;

public class MyReceiver extends BroadcastReceiver {
    public MyReceiver(){}
    @Override
    public void onReceive(Context context, Intent intent) {
        Toast.makeText(context,"Intent Detected.",Toast.LENGTH_LONG).show();
    }
}
```

修改新建工程的 res/layout 目录下的布局文件 activity_main.xml，修改代码如下：

```xml
<?xml version="1.0" encoding="utf-8"?>
<RelativeLayout xmlns:android="http://schemas.android.com/
        apk/res/android"
    xmlns:tools="http://schemas.android.com/tools" android:layout_width=
        "match_parent"
    android:layout_height="match_parent" android:paddingLeft="@dimen/
        activity_horizontal_margin"
    android:paddingRight="@dimen/activity_horizontal_margin"
    android:paddingTop="@dimen/activity_vertical_margin"
    android:paddingBottom="@dimen/activity_vertical_margin" tools:context=
        ".MainActivity">

    <TextView
        android:id="@+id/textView1"
        android:layout_width="wrap_content"
        android:layout_height="wrap_content"
        android:text="Example of Broadcast"
        android:layout_alignParentTop="true"
        android:layout_centerHorizontal="true"
        android:textSize="30dp" />

    <ImageButton
        android:layout_width="wrap_content"
        android:layout_height="wrap_content"
        android:id="@+id/imageButton"
        android:src="@drawable/ic_launcher"
        android:layout_centerVertical="true"
        android:layout_centerHorizontal="true" />

    <Button
        android:layout_width="wrap_content"
        android:layout_height="wrap_content"
        android:id="@+id/button"
```

```xml
            android:text="Broadcast Intent"
            android:onClick="broadcastIntent"
            android:layout_below="@+id/imageButton"
            android:layout_alignEnd="@+id/imageButton"
            android:layout_alignStart="@+id/imageButton" />
</RelativeLayout>
```

修改工程 java/com.example.broadcastcustom 目录下的 java 文件 MainActivity.java，修改代码如下：

```java
package com.example.broadcastcustom;

import android.app.Activity;
import android.content.Intent;
import android.os.Bundle;
import android.view.View;
import android.widget.Button;

public class MainActivity extends Activity {

    @Override
    protected void onCreate(Bundle savedInstanceState) {
        super.onCreate(savedInstanceState);
        setContentView(R.layout.activity_main);
        Button send= (Button)findViewById(R.id.button);
        send.setOnClickListener(new View.OnClickListener() {
            @Override
            public void onClick(View view) {
                Intent intent = new Intent();
                intent.setAction("com.example.action.DEMO_BROADCAST");
                sendBroadcast(intent);
            }
        });
    }

}
```

修改工程 manifests 目录下的配置文件 AndroidManifest.xml，修改代码如下：

```xml
<?xml version="1.0" encoding="utf-8"?>
<manifest xmlns:android="http://schemas.android.com/apk/res/android"
    package="com.example.broadcastcustom" >

    <application
        android:allowBackup="true"
        android:icon="@mipmap/ic_launcher"
        android:label="@string/app_name"
        android:supportsRtl="true"
        android:theme="@style/AppTheme" >
```

```xml
<activity android:name=".MainActivity" >
    <intent-filter>
        <action android:name="android.intent.action.MAIN" />

        <category android:name="android.intent.category.LAUNCHER" />
    </intent-filter>
</activity>

<receiver android:name=".MyReceiver" >
    <intent-filter android:priority="20" >
        <action android:name="com.example.action.DEMO_BROADCAST" >
        </action>
    </intent-filter>
</receiver>

    </application>

</manifest>
```

在 Genymotion 中运行，单击"Broadcast intent"按钮，效果如图 7.2 所示。

图 7.2　监听用户自定义 Broadcast Intent 消息实例运行效果

【例 7.2】　监听系统广播消息实例

这个例子将通过几个简单的步骤展示如何创建一个 Android Broadcast Receiver 用于监听手机电池状态的系统广播消息，如表 7.3 所示。

表 7.3 手机电池状态案例实例步骤

步骤	描述
1	打开 Android Studio 创建一个 Android 应用，Application name（应用名）取名为 BroadcastSystem，Company Domain（公司域）取名为 Example.com
2	Minimum SDK 选择 API 18: Android 4.3（Jelly Bean）
3	选择 Empty Activity
4	不需要修改 Activity Name，使用默认值，单击"Finish"按钮（完成）
5	右键单击工程的 com.example.broadcastcustom 目录，选择"New"→"Other"→"Broadcast Receiver"，新建一个 Broadcast Receiver 类取名 MyReceiver.java。该文件将定义一个 Broadcast Receiver
6	使用<receiver.../>标签在 AndroidManifest.xml 文件中注册 Broadcast Receiver，并用<intent-filter.../>标签设置 Broadcast Receiver 的过滤器
7	启动 Genymotion 模拟器，然后在 Android 工程中做如下代码修改

修改工程 java/com.example.broadcastsystem 目录下的服务类 java 文件 MyReceiver.java，修改代码如下：

```java
package com.example.broadcastsystem;

import android.content.BroadcastReceiver;
import android.content.Context;
import android.content.Intent;
import android.os.Bundle;
import android.widget.Toast;

public class MyReceiver extends BroadcastReceiver {
    @Override
    public void onReceive(Context context, Intent intent) {
        if (Intent.ACTION_BATTERY_OKAY.equals(intent.getAction())) {
            Toast.makeText(context, "电量已恢复,可以使用!", Toast.LENGTH_LONG).show();
        }
        if (Intent.ACTION_BATTERY_LOW.equals(intent.getAction())) {
            Toast.makeText(context, "电量过低,请尽快充电! ", Toast.LENGTH_LONG).show();
        }
        if (Intent.ACTION_BATTERY_CHANGED.equals(intent.getAction())) {
            Bundle bundle = intent.getExtras();
            //获取当前电量
            int current = bundle.getInt("level");
            //获取总电量
            int total = bundle.getInt("scale");
            StringBuffer sb = new StringBuffer();
            sb.append("当前电量为: " + current * 100 / total + "%" + "  ");
            //如果当前电量小于总电量的 15%
            if (current * 1.0 / total < 0.15) {
                sb.append("电量过低,请尽快充电! ");
            } else {
                sb.append("电量足够,请放心使用! ");
            }
            Toast.makeText(context, sb.toString(), Toast.LENGTH_LONG).show();
```

        }
    }
}

修改工程 manifests 目录下的配置文件 AndroidManifest.xml，修改代码如下：

```xml
<?xml version="1.0" encoding="utf-8"?>
<manifest xmlns:android="http://schemas.android.com/apk/res/android"
    package="com.example.broadcastsystem" >
    <uses-permission android:name="android.permission.BATTERY_STATS"/>
    <application
        android:allowBackup="true"
        android:icon="@mipmap/ic_launcher"
        android:label="@string/app_name"
        android:supportsRtl="true"
        android:theme="@style/AppTheme" >

        <activity android:name=".MainActivity" >
            <intent-filter>
                <action android:name="android.intent.action.MAIN" />

                <category android:name="android.intent.category.LAUNCHER" />
            </intent-filter>
        </activity>

        <receiver android:name=".MyReceiver" >
            <intent-filter>
                <action android:name="android.intent.action.BATTERY_CHANGED" />
                <action android:name="android.intent.action.BATTERY_OKAY"/>
                <action android:name="android.intent.action.BATTERY_LOW"/>
            </intent-filter>
        </receiver>
    </application>

</manifest>
```

在 Genymotion 模拟器中单击电池图标 ▯，弹出如图 7.3 所示的电池调节对话框。

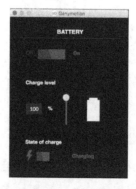

图 7.3　Genymotion 电池调节对话框

如图 7.3 所示，调节 Charge level（电池量）百分比，当电量低于 15%时，将会弹出"电量过低，请尽快充电！"的提示信息，如图 7.4 所示。

图 7.4　监听系统广播消息实例运行效果

这里，我们只演示了电量低的情况，读者可以调整电量测试其他几种电池状态的运行提示。

注意：例 7.2 在模拟器中运行的时候，可能会有延时的情况出现。请多尝试几次或在真机中运行。

第 8 章 ContentProvider 实现数据共享

当系统中部署越来越多的 Android 应用后，有时就需要在不同的应用之间共享数据。比如，现在有一个短信接收应用，用户想把接收到的短信的发送人添加到联系人管理应用中，就需要在不同应用之间共享数据。对于这种需求，操作方法有很多种，比如 SharedPreferences、文件或数据库等直接去操作另一个应用程序所记录的方式。但这种方式显得太杂乱，因为不同的应用程序记录数据的方式差别太大，不利于应用程序之间进行数据交换。

8.1 ContentProvider 概述

为了在应用程序之间交换数据，Android 提供了 ContentProvider。ContentProvider 是不同应用程序之间进行数据交换的标准 API，当一个应用程序需要把自己的数据交给其他应用程序使用时，该应用程序就可通过 ContentProvider 来实现；其他应用程序可以通过 ContentResolver 来操作 ContentProvider 提供的数据。ContentProvider 也可以把数据保存在数据库、文件甚至网络中。ContentProvider 工作模式如图 8.1 所示。

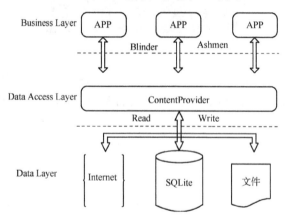

图 8.1 ContentProvider 运行模式

一旦某个应用程序通过 ContentProvider 共享了自己的数据操作接口，那么不管该应用程序是否启动，其他应用程序都可以通过该接口来操作共享出来的内部数据。ContentProvider 提供了 insert()（增加数据）、update()（修改数据）、delete()（删除数据）和 query()（查询数据）等数据操作方法。

创建一个自定义的 ContentProvider 类可以通过继承 Android 提供的 ContentProvider 基类来实现，例如以下代码：

```
public class My Application extends ContentProvider {

}
```

ContentProvider 也是 Android 应用的四大组件之一，它与 Activity、Service 和 BroadcastReceiver 相似，都需要在 AndroiManifest.xml 文件中进行注册，例如以下的注册代码：

```
<provider
    android:name=".StudentsProvider"
    android:authorities="com.example.provider.college"
    android:enabled="true"
    android:exported="true" >
</provider>
```

从上面的注册代码可以看出，注册 ContentProvider 时通常指定如表 8.1 所示的属性。

表 8.1 注册 ContentProvider 时的属性

属性	描述
name	指定该 ContentProvider 的实现类的类名
authorities	指定该 ContentProvider 对应的 URI
enabled	指定该 ContentProvider 是否可用
exported	指定该 ContentProvider 是否允许其他应用调用

8.2 URI 简介

每个 ContentProvider 都需要提供公共的 URI 来唯一标识其数据集。管理多个数据集的 ContentProvider 为每个数据集提供了单独的 URI。URI 的基本格式如下：

<prefix>://<authority>/<data_type>/<id>

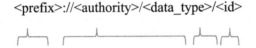

例如 content:// com.example.provider.college /people/2

表 8.2 详细介绍了每个组成部分。

表 8.2 URI 的组成部分详解

组成部分	描述
<prefix>	标准的前缀，始终设为 content://
<authority>	指定 ContentProvider 的名称。例如 contacts 或 browser 等（系统 ContentProvider）。对于第三方应用，该部分应该是完整的 ContentProvider 类名，例如 com.example.provider.college
<data_type>	指定 ContentProvider 的被请求数据的路径。例如用户需要从 contacts 这个系统的 ContentProvider 中查询所有联系人，那么 URI 为 content://contacts/people
<id>	被请求的特定记录的 id 值。例如用户需要从 contacts 这个系统的 ContentProvider 中查询第 5 个联系人，那么 URI 为 content://contacts/people/5

8.3 创建 ContentProvider

创建 ContentProvider 只需要如下两步：
- 像前面提到的一样，创建一个自定义的 ContentProvider 类可以通过继承 Android 提供的 ContentProvider 基类来实现。
- 在 AndroidManifest.xml 文件中注册该 ContentProvider，指定 android:authorities 属性：也就是为 ContentProvider 绑定一个 URI。

在上面两步中，自定义 ContentProvider 子类实现的 insert()（增加数据）、update()（修改数据）、delete()（删除数据）和 query()（查询数据）方法并不是给该应用本身调用的，而是供其他应用来调用的。当其他应用通过 ContentResolver 调用 insert()（增加数据）、update()（修改数据）、delete()（删除数据）和 query()（查询数据）方法执行数据访问时，实际上就是调用指定 URI 对应的 ContentProvider 的 insert()（增加数据）、update()（修改数据）、delete()（删除数据）和 query()（查询数据）方法。

如何重写 ContentProvider 的 insert()（增加数据）、update()（修改数据）、delete()（删除数据）和 query()（查询数据）等方法，完全取决于程序员。下面通过图 8.2 和表 8.3 对 ContentProvider 的需要重写的方法进行简单描述。

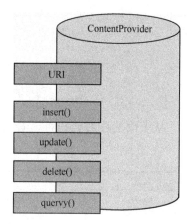

图 8.2　ContentProvider 子类的结构

表 8.3　ContentProvider 子类中的重写方法

重写方法	描述
onCreate()	当 ContentProvider 开始时，该方法被调用
query()	该方法接收客户端查询请求，结果返回一个 Cursor 对象
insert()	该方法为 ContentProvider 添加一条新记录
delete()	该方法从 ContentProvider 中删除一条记录
update()	该方法修改一条 ContentProvider 记录
getType()	该方法将返回给定 URI 指向数据的 MIME 类型

【例 8.1】　ContentProvider 实例

这个例子将通过几个简单的步骤展示如何创建一个 ContentProvider，如表 8.4 所示。

表 8.4　创建 ContentProvider 案例实例步骤

步骤	描述
1	打开 Android Studio 创建一个 Android 应用，Application name（应用名）取名为 ContentProvider，Company Domain（公司域）取名为 Example.com
2	Minimum SDK 选择 API 18: Android 4.3 (Jelly Bean)
3	选择 Empty Activity
4	不需要修改 Activity Name，使用默认值，单击"Finish"按钮（完成）
5	右键单击工程的 com.example.contentprovider 目录，选择"New"→"Other"→"Content Provider"，新建一个服务类取名 StudentsProvider.java。该文件将定义 ContentProvider 和相关方法
6	将工程 res/mipmap 目录下的 ic_launcher.png 图片文件复制、粘贴到 res/drawable 目录下
7	在工程中找到 ras/layout 目录中的 activity_main.xml 文件，在其中添加一个 textView（文本框）、一个 imageButton（图像按钮）、两个 editText（编辑框）（Name 和 Grade）和两个 Button（普通按钮）控件（Add Name 和 Retrive student），按图 8.3 所示的运行效果位置进行摆放
8	按照如下代码修改 java/com.example.contentprovider 目录下 java 文件 MainActivity.java 代码，为其添加 onClickAddName() 和 onClickRetrieveStudents()方法
9	使用<provider.../>标签在 AndroidManifest.xml 文件中注册 ContentProvider
10	启动 Genymotion 模拟器，然后在 Android 工程中做如下代码修改

修改工程的 res/layout 目录下的布局文件 activity_main.xml，修改代码如下：

```
<?xml version="1.0" encoding="utf-8"?>
```

```xml
<RelativeLayout xmlns:android="http://schemas.android.com/
        apk/res/android"
    xmlns:tools="http://schemas.android.com/tools" android:layout_width=
        "match_parent"
    android:layout_height="match_parent" android:paddingLeft=
        "@dimen/activity_horizontal_margin"
    android:paddingRight="@dimen/activity_horizontal_margin"
    android:paddingTop="@dimen/activity_vertical_margin"
    android:paddingBottom="@dimen/activity_vertical_margin" tools:context=
        ".MainActivity">

    <TextView
        android:id="@+id/textView1"
        android:layout_width="wrap_content"
        android:layout_height="wrap_content"
        android:text="Content provider"
        android:layout_alignParentTop="true"
        android:layout_centerHorizontal="true"
        android:textSize="30dp" />

    <ImageButton
        android:layout_width="wrap_content"
        android:layout_height="wrap_content"
        android:id="@+id/imageButton"
        android:src="@drawable/ic_launcher"
        android:layout_centerHorizontal="true"
        android:layout_below="@+id/textView1" />

    <Button
        android:layout_width="wrap_content"
        android:layout_height="wrap_content"
        android:id="@+id/button2"
        android:text="Add Name"
        android:layout_below="@+id/editText3"
        android:onClick="onClickAddName"
        android:layout_alignEnd="@+id/editText3"
        android:layout_alignStart="@+id/editText3" />

    <EditText
        android:layout_width="wrap_content"
        android:layout_height="wrap_content"
        android:id="@+id/editText2"
        android:hint="Name"
        android:textColorHint="@android:color/holo_blue_light"
        android:layout_below="@+id/imageButton"
        android:layout_alignEnd="@+id/editText3"
        android:layout_alignStart="@+id/editText3" />
```

```xml
<EditText
    android:layout_width="wrap_content"
    android:layout_height="wrap_content"
    android:id="@+id/editText3"
    android:hint="Grade"
    android:textColorHint="@android:color/holo_blue_bright"
    android:layout_below="@+id/editText2"
    android:layout_alignEnd="@+id/imageButton"
    android:layout_alignStart="@+id/imageButton" />

<Button
    android:layout_width="wrap_content"
    android:layout_height="wrap_content"
    android:text="Retrive student"
    android:id="@+id/button"
    android:layout_below="@+id/button2"
    android:layout_alignRight="@+id/editText3"
    android:layout_alignEnd="@+id/editText3"
    android:layout_alignLeft="@+id/button2"
    android:layout_alignStart="@+id/button2"
    android:onClick="onClickRetrieveStudents"/>
</RelativeLayout>
```

修改工程 java/com.example.contentprovider 目录下的服务类 java 文件 StudentsProvider.java，修改代码如下：

```java
package com.example.contentprovider;

import android.content.ContentProvider;
import android.content.ContentUris;
import android.content.ContentValues;
import android.content.Context;
import android.content.UriMatcher;
import android.database.Cursor;
import android.database.SQLException;
import android.database.sqlite.SQLiteDatabase;
import android.database.sqlite.SQLiteOpenHelper;
import android.database.sqlite.SQLiteQueryBuilder;
import android.net.Uri;
import android.text.TextUtils;

import java.util.HashMap;

public class StudentsProvider extends ContentProvider {
    static final String PROVIDER_NAME = "com.example.provider.College";
    static final String URL = "content://" + PROVIDER_NAME + "/students";
    static final Uri CONTENT_URI = Uri.parse(URL);
```

```java
static final String _ID = "_id";
static final String NAME = "name";
static final String GRADE = "grade";

private static HashMap<String, String> STUDENTS_PROJECTION_MAP;

static final int STUDENTS = 1;
static final int STUDENT_ID = 2;

static final UriMatcher uriMatcher;
static{
    uriMatcher = new UriMatcher(UriMatcher.NO_MATCH);
    uriMatcher.addURI(PROVIDER_NAME, "students", STUDENTS);
    uriMatcher.addURI(PROVIDER_NAME, "students/#", STUDENT_ID);
}

/**
 * Database specific constant declarations
 */
private SQLiteDatabase db;
static final String DATABASE_NAME = "College";
static final String STUDENTS_TABLE_NAME = "students";
static final int DATABASE_VERSION = 1;
static final String CREATE_DB_TABLE =
        " CREATE TABLE " + STUDENTS_TABLE_NAME +
            " (_id INTEGER PRIMARY KEY AUTOINCREMENT, " +
            " name TEXT NOT NULL, " +
            " grade TEXT NOT NULL);";

/**
 * Helper class that actually creates and manages
 * the provider's underlying data repository.
 */
private static class DatabaseHelper extends SQLiteOpenHelper {
    DatabaseHelper(Context context){
        super(context, DATABASE_NAME, null, DATABASE_VERSION);
    }

    @Override
    public void onCreate(SQLiteDatabase db)
    {
        db.execSQL(CREATE_DB_TABLE);
    }

    @Override
    public void onUpgrade(SQLiteDatabase db, int oldVersion, int newVersion) {
```

```java
            db.execSQL("DROP TABLE IF EXISTS " + STUDENTS_TABLE_NAME);
            onCreate(db);
        }
    }

    @Override
    public boolean onCreate() {
        Context context = getContext();
        DatabaseHelper dbHelper = new DatabaseHelper(context);

        /**
         * Create a write able database which will trigger its
         * creation if it doesn't already exist.
         */
        db = dbHelper.getWritableDatabase();
        return (db == null)? false:true;
    }

    @Override
    public Uri insert(Uri uri, ContentValues values) {
        /**
         * Add a new student record
         */
        long rowID = db.insert( STUDENTS_TABLE_NAME, "", values);

        /**
         * If record is added successfully
         */

        if (rowID > 0)
        {
            Uri _uri = ContentUris.withAppendedId(CONTENT_URI, rowID);
            getContext().getContentResolver().notifyChange(_uri, null);
            return _uri;
        }
        throw new SQLException("Failed to add a record into " + uri);
    }

    @Override
    public Cursor query(Uri uri, String[] projection, String selection,
        String[] selectionArgs, String sortOrder) {
        SQLiteQueryBuilder qb = new SQLiteQueryBuilder();
        qb.setTables(STUDENTS_TABLE_NAME);

        switch (uriMatcher.match(uri)) {
            case STUDENTS:
                qb.setProjectionMap(STUDENTS_PROJECTION_MAP);
```

```java
            break;

        case STUDENT_ID:
            qb.appendWhere( _ID + "=" + uri.getPathSegments().get(1));
            break;

        default:
            throw new IllegalArgumentException("Unknown URI " + uri);
    }

    if (sortOrder == null || sortOrder == ""){
        /**
         * By default sort on student names
         */
        sortOrder = NAME;
    }
    Cursor c = qb.query(db,    projection,   selection, selectionArgs,
            null, null, sortOrder);

    /**
     * register to watch a content URI for changes
     */
    c.setNotificationUri(getContext().getContentResolver(), uri);
    return c;
}

@Override
public int delete(Uri uri, String selection, String[] selectionArgs) {
    int count = 0;

    switch (uriMatcher.match(uri)){
        case STUDENTS:
            count = db.delete(STUDENTS_TABLE_NAME, selection, selectionArgs);
            break;

        case STUDENT_ID:
            String id = uri.getPathSegments().get(1);
            count = db.delete( STUDENTS_TABLE_NAME, _ID +  " = " + id +
                    (!TextUtils.isEmpty(selection) ? " AND (" + selection
                    + ')' : ""), selectionArgs);
            break;

        default:
            throw new IllegalArgumentException("Unknown URI " + uri);
    }

    getContext().getContentResolver().notifyChange(uri, null);
```

```java
            return count;
        }

        @Override
        public int update(Uri uri, ContentValues values, String selection,
                          String[] selectionArgs) {
            int count = 0;

            switch (uriMatcher.match(uri)){
                case STUDENTS:
                    count = db.update(STUDENTS_TABLE_NAME, values, selection,
                                      selectionArgs);
                    break;

                case STUDENT_ID:
                    count = db.update(STUDENTS_TABLE_NAME, values, _ID + " = "
                            + uri.getPathSegments().get(1) +
                            (!TextUtils.isEmpty(selection) ? " AND (" +selection
                            + ')' : ""), selectionArgs);
                    break;

                default:
                    throw new IllegalArgumentException("Unknown URI " + uri );
            }
            getContext().getContentResolver().notifyChange(uri, null);
            return count;
        }

        @Override
        public String getType(Uri uri) {
            switch (uriMatcher.match(uri)){
                /**
                 * Get all student records
                 */
                case STUDENTS:
                    return "vnd.android.cursor.dir/vnd.example.students";

                /**
                 * Get a particular student
                 */
                case STUDENT_ID:
                    return "vnd.android.cursor.item/vnd.example.students";

                default:
                    throw new IllegalArgumentException("Unsupported URI: " + uri);
            }
        }
    }
```

修改工程 java/com.example.contentprovider 目录下的 java 文件 MainActivity.java，修改代码如下：

```java
package com.example.contentprovider;

import android.app.Activity;
import android.content.ContentValues;
import android.database.Cursor;
import android.net.Uri;
import android.os.Bundle;
import android.view.View;
import android.widget.EditText;
import android.widget.Toast;

public class MainActivity extends Activity {

    @Override
    protected void onCreate(Bundle savedInstanceState) {
        super.onCreate(savedInstanceState);
        setContentView(R.layout.activity_main);
    }
    public void onClickAddName(View view) {
        //Add a new student record
        ContentValues values = new ContentValues();

        values.put(StudentsProvider.NAME,
                ((EditText)findViewById(R.id.editText2)).getText().toString());

        values.put(StudentsProvider.GRADE,
                ((EditText)findViewById(R.id.editText3)).getText().toString());

        Uri uri = getContentResolver().insert(
                StudentsProvider.CONTENT_URI, values);

        Toast.makeText(getBaseContext(),
                uri.toString(), Toast.LENGTH_LONG).show();
    }

    public void onClickRetrieveStudents(View view) {

        //Retrieve student records
        String URL = "content://com.example.provider.College/students";

        Uri students = Uri.parse(URL);
        Cursor c = managedQuery(students, null, null, null, "name");

        if (c.moveToFirst()) {
            do{
```

```
                Toast.makeText(this,
                    c.getString(c.getColumnIndex(StudentsProvider._ID)) +
                        ", " + c.getString(c.getColumnIndex
                        (StudentsProvider.NAME)) +
                        ", " + c.getString(c.getColumnIndex
                        (StudentsProvider.GRADE)),
                    Toast.LENGTH_SHORT).show();
            } while (c.moveToNext());
        }
    }
}
```

修改工程 manifests 目录下的配置文件 AndroidManifest.xml，修改代码如下：

```xml
<?xml version="1.0" encoding="utf-8"?>
<manifest xmlns:android="http://schemas.android.com/apk/res/android"
    package="com.example.contentprovider" >

    <application
        android:allowBackup="true"
        android:icon="@mipmap/ic_launcher"
        android:label="@string/app_name"
        android:supportsRtl="true"
        android:theme="@style/AppTheme" >
        <activity android:name=".MainActivity" >
            <intent-filter>
                <action android:name="android.intent.action.MAIN" />

                <category android:name="android.intent.category.LAUNCHER" />
            </intent-filter>
        </activity>

        <provider
            android:name=".StudentsProvider"
            android:authorities="com.example.provider.College"
            android:enabled="true"
            android:exported="true" >
        </provider>
    </application>

</manifest>
```

在 Genymotion 中运行，然后在"Name"编辑框中输入名字，在"Grade"编辑框中输入年级，单击"Add Name"按钮，效果如图 8.3 所示。

用同样的方式连续输入多个学生信息，学生信息将会通过 insert()方法保存到 ContentProvider 的数据库中，然后单击"Retrive student"按钮。数据库中保存的信息将会通过 query()方法返回并依次显示在屏幕中，运行效果如图 8.4 所示。

第 8 章 ContentProvider 实现数据共享

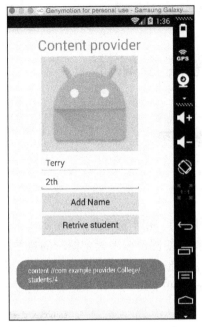

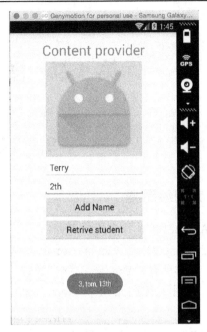

图 8.3　创建 ContentProvider 实例运行效果 1　　　图 8.4　创建 ContentProvider 实例运行效果 2

第 9 章 图形图像与多媒体

9.1 基础绘图

9.1.1 常用绘图类

在 Android 中，许多组件都是 View 的子类，所以在绘图时应当继承 View 类，并重写其中的指定方法 onDraw(Canvas canvas)。当你想展示特定的图形或者设定图形的特效时，需要具备 4 个基本组件：Bitmap 类保存像素、Canvas 类保存绘图的回调结果（在 bitmap 中要绘制的内容）、通用绘图图元（例如 Rect、Path、text、Bitmap）和 Print 类（用来设置颜色以及绘图的风格）。

Android 中的常用绘图类包括以下几个。

（1）android.graphics.Bitmap：代表一张位图，所包含的图片可以由程序创建或者来自文件。

（2）android.grapics.Canvas：表示画布，可以通过设置画布的属性，如颜色或者尺寸等，画出你想画的东西。

（3）android.grapics.Paint：代表画笔，用来描述绘制图形的风格，例如设置颜色、线宽、透明度等属性，使用该类时，可以通过该类提供的构造方法创建对象。

在一般的绘图操作中，Canvas 类提供了一些常用的绘图方法，比如画圆形、正方形、长方形、椭圆形，这些方法可以直接画出图形。Canvas 提供了如表 9.1 所示的绘图方法。

表 9.1 Canvas 常用绘图方法

返回类型	方法	简述
boolean	clipRegion(Region region)	裁剪指定区域
	clipRect(int left, int top, int right, int bottom)	裁剪一个矩形区域
void	drawPoint(float x, float y, Paint paint)	绘制一个点
	drawPoints(float[] pts, int offset, int count, Paint paint)	绘制多个点
	drawLine(float startX, float startY, float stopX, float stopY, Paint paint)	绘制一条线
	drawLines(float[] pts, int offset, int count, Paint paint)	绘制多条线
	drawText(String text, float x, float y, Paint paint)	绘制字符串
	drawRect(float left, float top, float right, float bottom, Paint paint)	绘制矩形
	drawRoundRect(RectF rect, float rx, float ry, Paint paint)	绘制圆角矩形
	drawOval(float left, float top, float right, float bottom, Paint paint)	绘制椭圆
	drawArc(RectF oval, float startAngle, float sweepAngle, boolean useCenter, Paint paint)	绘制弧
	drawPaint(Paint paint)	在画布上绘满位图
	drawPath(Path path, Paint paint)	沿着指定 Path 绘制任意形状
	drawBitmap(Bitmap bitmap, float left, float top, Paint paint)	在指定坐标绘制位图
	drawBitmap(Bitmap bitmap, Rect src, Rect dst, Paint paint)	在指定坐标绘制从源位图中"挖取"的一块

续表

返回类型	方法	简述
void	drawCircle(float cx, float cy, float radius, Paint paint)	在指定坐标绘制圆形
	drawColor(int color)	用指定的颜色填充满位图
	drawRGB(int r, int g, int b)	用指定的颜色 RGB 填充满位图
	drawARGB(int a, int r, int g, int b)	用指定的颜色 ARGB 填充满位图

在使用 Canvas 类进行绘图时，还需要结合 Paint 类，Paint 代表了 Canvas 上的画笔，用于绘图属性的设置。Paint 提供了如表 9.2 所示的绘图方法。

表 9.2 Paint 常用绘图方法

返回类型	方法	简述
boolean	setARGB(int a,int r,int g,int b)	设置颜色，参数值分别表示透明度、红色、绿色和蓝色值
	setAlpha(int a)	设置透明度
	setColor(int color)	设置颜色
	setAntiAlias(boolean aa)	设置是否使用抗锯齿功能
	setDither(boolean dither)	设置是否使用图像抖动处理
	setPathEffect(PathEffect effect)	设置绘制路径时的路径效果
	setShader(Shader shader)	设置渐变
	setShadowLayer(float radius,float dx,float dy,int color)	设置阴影
	setStyle(Paint.Style style)	设置画笔的样式风格
	setStrokeCap(Paint.Cap cap)	设置笔刷的图形样式
	setStrokeWidth(float width)	设置笔刷的宽度
	setSrokeJoin(Paint.Join join)	设置绘制时各图形的连接方式
	setXfermode(Xfermode xfermode)	设置图形重叠时的处理方式
	setTextAlign(Paint.Align align)	设置绘制文字的对齐方式
	setTextSize(float textSize)	设置绘制文字的字号大小
	setUnderlineText(boolean underlineText)	设置带有下画线的文字效果
	setStrikeThruText(boolean strikeThruText)	设置带有删除线的文字效果

9.1.2 绘制 2D 图形

在重写 onDraw()方法时，可以通过调用一些方法绘制基础的 2D 图形。下面详细介绍 Android 中绘制简单的 2D 图形用到的方法。

1）点

drawPoint(float x,float y,Paint paint)：在指定坐标绘制一个点。

2）直线

drawLine(float startx,float starty,loat stopx,float stopy,Paint paint)：在点（startx,starty）和点（stopx,stopy）之间绘制一条直线。

3）矩形

drawRect(float x1,float y1,float x2,float y2,Paint paint)：绘制左上角顶点为（x1,y1),右下角顶点为(x2,y2)的矩形。

4) 多边形

drawVertices(VertexMode mode,int count,float[] pts,0,null,0,null,0,null,0,0,Paint paint)：绘制一个多边形，其中 count 为坐标的个数，等于 pts 的大小。

5) 圆

drawCircle(float cx,float cy,float r,Paint paint)：绘制一个以（cx,cy）为圆心、r 为半径的圆。

6) 弧线

drawArc(Rect rect,float startAngle,float sweepAngle,bool useCenter,Paint paint)：在矩形 rect 内部（此矩形并不会绘制）绘制一个起始角度为 startAngle、结束角度为 sweepAngle 的弧，useCenter 将决定这段弧是否会显示起点与终点的连线。

7) 文字

drawText(String s, float x, float y, Paint paint)：在（x,y）处绘制字符串 s。

【例 9.1】 Canvas 类与 Paint 类实例

下面使用 Canvas 类与 Paint 类中的一些常用的方法，绘制一个 Android 机器人图标，具体实现方法如下：

```
定义MyView类，此类继承View类
package com.example.canvaspaint;
import android.content.Context;
import android.graphics.Canvas;
import android.graphics.Color;
import android.graphics.Paint;
import android.graphics.Rect;
import android.graphics.RectF;
import android.util.AttributeSet;
import android.view.View;
public class MyView extends View {
    public MyView(Context context,AttributeSet attrs)          //继承View类
    {
        super(context,attrs);
    }
    @Override
    protected void onDraw(Canvas canvas) {
        Paint paint=new Paint();                               //采用默认设置创建一个画笔
        paint.setAntiAlias(true);                              //使用抗锯齿功能
        paint.setColor(0xFF78C257);                            //设置画笔的颜色为绿色
        //绘制机器人的头
        RectF rectf_head = new RectF(110, 30, 200, 120);
        canvas.drawArc(rectf_head, -10, -160, false, paint);   //绘制弧
        //绘制眼睛
        paint.setColor(Color.WHITE);                           //设置画笔的颜色为白色
        canvas.drawCircle(135, 53, 4, paint);                  //绘制圆
        canvas.drawCircle(175, 53, 4, paint);                  //绘制圆
        //绘制天线
        paint.setColor(0xFF78C257);                            //设置画笔的颜色为绿色
        paint.setStrokeWidth(8);                               //设置笔触的宽度
```

```java
        canvas.drawLine(125, 20, 135, 35, paint);        //绘制线
        canvas.drawLine(185, 20, 175, 35, paint);        //绘制线
        //绘制身体
        canvas.drawRect(110, 75, 200, 150, paint);       //绘制矩形
        RectF rectf_body = new RectF(110,140,200,160);
        canvas.drawRoundRect(rectf_body, 10, 10, paint); //绘制圆角矩形
        //绘制胳膊
        RectF rectf_arm = new RectF(85,75,105,140);
        canvas.drawRoundRect(rectf_arm, 10, 10, paint);  //绘制左侧的胳膊
        rectf_arm.offset(120, 0);                        //设置在X轴上偏移120像素
        canvas.drawRoundRect(rectf_arm, 10, 10, paint);  //绘制右侧的胳膊
        //绘制腿
        RectF rectf_leg = new RectF(125,150,145,200);
        canvas.drawRoundRect(rectf_leg, 10, 10, paint);  //绘制左侧的腿
        rectf_leg.offset(40, 0);                         //设置在X轴上偏移40像素
        canvas.drawRoundRect(rectf_leg, 10, 10, paint);  //绘制右侧的腿
        //写文字
        paint.setTextSize(50);                           //设置字体大小
        paint.setColor(Color.GRAY);                      //设置画笔颜色为灰色
        canvas.drawText("ANDROID",45,260,paint);         //绘制文字
        super.onDraw(canvas);
    }
}
```

然后，在布局管理器中使用定义的 MyView 组件：

```xml
<?xml version="1.0" encoding="utf-8"?>
<RelativeLayout xmlns:android="http://schemas.android.com/
        apk/res/android"
    xmlns:tools="http://schemas.android.com/tools"
    xmlns:app=http://schemas.android.com/apk/res-auto android:layout_width=
        "match_parent"
    android:layout_height="match_parent" android:paddingLeft="@dimen/
        activity_horizontal_margin"
    android:paddingRight="@dimen/activity_horizontal_margin"
    android:paddingTop="@dimen/activity_vertical_margin"
    android:paddingBottom="@dimen/activity_vertical_margin"
    app:layout_behavior="@string/appbar_scrolling_view_behavior"
    tools:showIn="@layout/activity_main" tools:context=".MainActivity">
    <com.example.canvaspaint
        android:layout_width="fill_parent"
        android:layout_height="fill_parent" />
</RelativeLayout>
```

在 Genymotion 中运行程序的效果如图 9.1 所示。

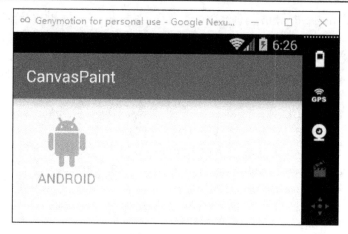

图 9.1　基础绘图

9.2　位图操作

学会如何绘制图像后，懂得图像处理也是至关重要的。Bitmap 类，即位图，是 Android 中图像处理基础、重要的类之一，使用它可以实现图像文件信息的读取和写入，进行图像剪切、旋转、缩放等操作，并可按指定格式保存图像文件。Bitmap 类的常用方法如表 9.3 所示。

表 9.3　Bitmap 类的常用方法

返回类型	方法	简述
static Bitmap	creatBitmap(Bitmap source, int x, int y, int width, int height)	从源位图中创建新的 Bitmap 对象
	creatScaledBitmap(Bitmap src, int dstWidth, int dstheight, boolean filter)	对源位图进行缩放，构成新位图
	creatBitmap(int width, int height, Bitmap.Config config)	根据指定的数据、配置创建新位图
	creatBitmap(Bitmap source, int x, int y, int width, int height, Matrix m, boolean filter)	从源位图中创建新的 Bitmap 对象，并进行 Matrix 变换
final int	getHeight()	获取位图的高
	getWeight()	获取位图的宽

在 Android 中，Bitmap 比较特别，因为 Bitmap 类的构造函数是私有的，并不能实例化，而只能借助于 BitmapFactory 类获取文件资源，从而得到一个实例。BitmapFactory 类的常用绘图方法如表 9.4 所示。

表 9.4　BitmapFractory 类的常用绘图方法

返回类型	方法	简述
static Bitmap	decodeByteArray(byte[] data, int offset, int length)	根据指定的数据文件解码 Bitmap
	decodeFile(String pathName)	根据指定的路径创建 Bitmap
	decodeFileDescriptor(FileDescriptor fd)	根据指定的文件说明符解码 Bitmap
	decodeResource(Resources res, int id)	根据指定的资源创建 Bitmap
	decodeStream(InputStream is)	根据指定的输入流解码 Bitmap

从以上的方法中不难看出，BitmapFactory 相当于从一个指定的文件中获取并解析出 Bitmap。一般有两种读取 Bitmap 的方法。

方法一：以 R 文件的方式，我们一般把图片放在/res/drawable 目录下，所以直接在程序中通过该图片对应的资源号就可以获取图片对象，例如 res/drawable 下有 test.png 图片，代码如下：

```
Bitmap bitmap = BitmapFactory.decodeResource(this.getContext().getResorces,
R.drawable.test)
```

方法二：以文件流的方式，假设在 sdcard 下有 test.png 图片：

```
FileInputStream fis = new FileInputStream("/sdcard/test.png");
Bitmap bitmap = BitmapFactory.decodeStream(fis);
```

【例 9.2】 Bitmap 获取图片实例

下面例子展示了运用 Bitmap 获取图片并用按钮切换显示：

```
package com.example.bitmap;
import android.graphics.Bitmap;
import android.graphics.BitmapFactory;
import android.support.v7.app.AppCompatActivity;
import android.os.Bundle;
import android.view.View;
import android.widget.Button;
import android.widget.ImageView;
import import MainActivity import AppCompatActivity {
    ImageView img = null;
    Button but = null;
    int count = 0;
    int[] imgRes = {R.drawable.leaf1,R.drawable.leaf2,R.drawable.leaf3};
    @Override
    protected void onCreate(Bundle savedInstanceState) {
        super.onCreate(savedInstanceState);
        setContentView(R.layout.activity_main);
        Bitmap bm = BitmapFactory.decodeResource(this.getResources(),
                imgRes[count]);
        img = (ImageView)findViewById(R.id.imageView);
        but = (Button)findViewById(R.id.but_next);
        img.setImageBitmap(bm);
        but.setOnClickListener(new View.OnClickListener() {
            @Override
            public void onClick(View v) {
                count++;
                if(count==3)
                {count=0;}
                Bitmap bm = BitmapFactory.decodeResource(MainActivity.
                        this.getResources(),imgRes[count]);
                img.setImageBitmap(bm);
            }
        });
```

```
    }
}
```

本程序将保存在 res/drawable 中的图片资源提取出来,保存在 bitmap 中,并用按钮来切换资源图片。

在布局文件(activity_main.xml)中定义相关的组件:

```xml
<?xml version="1.0" encoding="utf-8"?>
<LinearLayout xmlns:android="http://schemas.android.com/
        apk/res/android"
    xmlns:tools="http://schemas.android.com/tools"
    tools:context=".MainActivity"
    android:layout_width="match_parent"
    android:layout_height="match_parent"
    android:orientation="vertical">
    <ImageView
        android:id="@+id/imageView"
        android:layout_width="fill_parent"
        android:layout_height="200dp" />
    <Button
        android:id="@+id/but_next"
        android:layout_width="wrap_content"
        android:layout_height="wrap_content"
        android:textSize="10dp"
        android:text="@string/next_pic"
        android:layout_gravity="center_horizontal"
        />
</LinearLayout>
```

在布局文件中定义了 ImageView 显示图片,以及一个按钮切换图片。在 Genymotion 中的运行效果如图 9.2 所示。

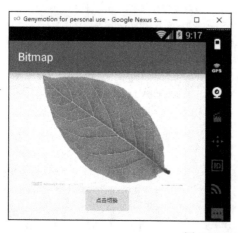

图 9.2　Bitmap 显示图片

但是在运行上述例子时,会出现一些卡顿的现象,甚至会出现 OOM(Out of Memory)内存溢出问题,这是因为图片较大。所以在实际运用时,一般会通过设置 BitmapFactory.Options 的一些参数来对图像加载做一些优化,代码如下:

```
BitmapFactory.Options opt = new BitmapFactory.Options();
opt.inSampleSize = 2;                    /*2倍缩放图片*/
opt.inDither=false;                      /*不进行图片抖动处理*/
opt.inPreferredConfig=null;              /*设置让解码器以最佳方式解码*/
```

在设置了几个参数之后，图片切换就十分流畅了。

9.3 Android 中的动画

在 Android 3.0 以前，Android 支持以下两种动画模式。

(1) Frame Animation(逐帧动画)：可以将一系列的图像依次显示，以模拟动画效果，就像 GIF 图一样。

(2) Tween Animation(补间动画)：对图形做基本动画处理，例如渐变、旋转、伸缩以及移动的操作。

之后，又添加了下述一种新的动画方式。

Property Animation(属性动画)：利用对象的属性变化形成动画效果，如按钮的位置和大小发生改变。

这三种动画在 SDK 中分别被称为 Drawable animation、View animation、Property animation。本节将详细介绍这三种动画的运用方式。

9.3.1 Frame Animation

Frame Animation，即逐帧动画，通过一系列 Drawable 控制依次显示动画过程中的每张静态图片，利用人眼"视觉暂留"的原理以达到动画效果。在 XML 文件中可以对这种动画进行配置，并将文件保存在 res/drawable 文件夹中。配置逐帧动画非常简单，在<animation-list.../>元素中逐个配置<item>元素即可定义动画的各个帧，制定各个帧的持续时间即可，相关的配置属性如表 9.5 所示。

表 9.5 逐帧动画的属性描述

属性	说明	参数说明
oneshot	循环属性	当设置为 true 时，动画只循环一次，否则一直循环
drawable	动画资源	工程目录下的动画资源文件
visible	drawable 是否初始可见	当设备为 true 时，动画初始可见，否则初始不可见
duration	动画持续时间	时间以毫秒为单位

【例 9.3】 利用 Frame Animation 实现缓冲圈效果的实例

首先，在 drawable 文件夹中定义 my_frameanimation.xml 文件如下：

```
<?xml version="1.0" encoding="utf-8"?>
<animation-list
    xmlns:android="http://schemas.android.com/apk/res/android"
    android:oneshot="false"
    >
    <item
        android:drawable="@drawable/pic1"
```

```xml
            android:duration="200"
            />
        <item
            android:drawable="@drawable/pic2"
            android:duration="200"
            />
        <item
            android:drawable="@drawable/pic3"
            android:duration="200"
            />
        <item
            android:drawable="@drawable/pic4"
            android:duration="200"
            />
        <item
            android:drawable="@drawable/pic5"
            android:duration="200"
            />
        <item
            android:drawable="@drawable/pic6"
            android:duration="200"
            />
        <item
            android:drawable="@drawable/pic7"
            android:duration="200"
            />
        <item
            android:drawable="@drawable/pic8"
            android:duration="200"
            />
</animation-list>
```

然后，在布局文件中添加 ImageView 组件：

```xml
<LinearLayout xmlns:android="http://schemas.android.com/apk/
        res/android"
    xmlns:tools="http://schemas.android.com/tools"
    android:id="@+id/linear"
    android:layout_width="fill_parent"
    android:layout_height="fill_parent"
android:orientation="vertical" >
    <ImageView
        android:id="@+id/imageView1"
        android:layout_width="wrap_content"
        android:layout_height="wrap_content"
        android:layout_gravity="center" />
</LinearLayout>
```

最后，在程序中调用该动画：

```java
package com.example.frameanimation;
import com.example. frameanimation.R;
import android.app.Activity;
import android.graphics.drawable.AnimationDrawable;
import android.os.Bundle;
import android.widget.ImageView;
public class MainActivity extends Activity {
    private ImageView mImg;
    private AnimationDrawable mAd;
    @Override
    protected void onCreate(Bundle savedInstanceState) {
        super.onCreate(savedInstanceState);
        setContentView(R.layout.activity_main);
        mImg = (ImageView) findViewById(R.id.imageView1);
        mImg.setBackgroundResource(R.drawable.my_frameanimation);
        mAd = (AnimationDrawable) MainActivity.this.mImg.getBackground();
        mAd.start();
    }
}
```

动画效果如图 9.3 所示。

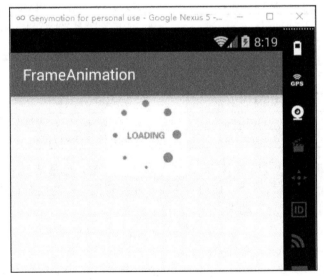

图 9.3　Frame Animation 效果

9.3.2　Tween Animation

Tween Animation，即补间动画，可以让某个控件出现旋转、渐变、移动、缩放。补间动画实现的原理比较简单，开发者不需定义每一帧的图像，只需要定义动画开始与结束的关键帧(首帧和尾帧)，同时制定动画执行的时间，其过程如图 9.4 所示。

Tween Animation 只能应用于 View 对象，而且只支持一部分属性，如支持缩放旋转而不支持背景颜色的改变等。

Tween Animation 主要由以下 4 种类型组成。

（1）alpha（android.view.animation.Alpha.Animation）：渐变动画，可实现 fadeIn 和 fadeOut 效果。

（2）rotate（android.view.animation.Rotate.Animation）：旋转动画，可实现移动控件的旋转效果。

（3）scale（android.view.animation.Scale.Animation）：缩放动画，可实现动态调整控件尺寸的效果。

（4）translate（android.view.animation.Translate.Animation）：位移动画，可实现控件水平、垂直位移的效果。

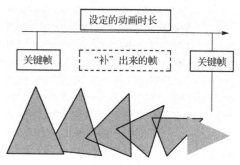

图 9.4　补间动画示意图

Android 中的 4 种 animation 类型在 Java 中对应相应的类：AlphaAnimation、ScaleAnimation、TranslateAnimation 及 RotateAnimation。可以调用这些类的相关方法来获取和操作相应的属性。这 4 种类型的部分属性是相同的，如表 9.6 所示。

表 9.6　Tween Animation 共同属性描述

属性	方法	简述
android:detachWallpaper	setDetachWallpaper(boolean)	当窗口在墙纸顶层时，窗口动画，墙纸不动画
android:duration	setDuration(long)	总计动画运行时间
android:fillAfter	setFillAfter(boolean)	其值为 true 时，该动画转化在动画结束后被应用
android:fillBefore	setFillBefore(boolean)	其值为 true 或 fillEnabled 为 false 时，该动画转化在动画开始前被应用
android:fillEnabled	setFillEnabled(boolean)	其值为 true 时，fillBefore 的值被考虑
android:interpolator	setInterpolator(Interpolator)	设置动画的 interpolator
android:repeatCount	setRepeatCount(int)	设置动画重复的次数
android:repeatMode	setRepeatMode(int)	设置动画重复的模式
android:startOffset	setStartOffset(long)	设置在动画运行前延迟数毫秒
android:zAdjustment	setZAdjustment(int)	设置动画的 Z Order 的改变

可以编码实现动画，也可以用 XML 文件形式定义动画。通常情况下，用 XML 文件形式定义动画，并放在/res/anim 目录下，这符合 MVC 开发规范，在需要时直接调用资源即可。下面介绍如何在 XML 文件中定义动画。

（1）新建 Android 项目。

（2）在 res 目录中新建 anim 文件夹。

（3）在 anim 目录中新建一个 new_anim.xml 的文件（注意：Android 项目中 XML 类型文件名必须都用小写）。

（4）在 new_anim.xml 中加入动画代码。

① alpha 渐变透明度动画效果的属性如表 9.7 所示。

表 9.7　alpha 属性描述

属性	说明	参数说明
fromAlpha	动画起始时透明度	以上值取为 0.0～1.0 之间的 float 数据类型的数字，0.0 表示完全透明，1.0 表示完全不透明
toAlpha	动画结束时透明度	
duration	动画持续时间	时间以毫秒为单位

例如，实现一个持续 1s 由 50%透明度到完全透明效果的动画，代码如下：

```
<?xml version="1.0" encoding="utf-8"?>
<set xmlns:android="http://schemas.android.com/apk/res/android" >
    <alpha
        android:duration="1000"
        android:fromAlpha="0.5"
        android:toAlpha="1.0" />
</set>
```

② rotate 画面移动旋转动画效果属性如表 9.8 所示。

表 9.8 rotate 属性描述

属性	说明	参数说明
fromDegrees	动画起始时物件的角度	当起始角度小于结束角度时，物件顺时针旋转
toDegrees	动画结束时物件旋转的角度	
pivotX	动画相对于物件的 X 坐标的开始位置	从 0%～100%中取值
PivotY	动画相对于物件的 Y 坐标的开始位置	50%为物件的 X 或 Y 方向坐标上的中点位置
duration	动画持续时间	时间以毫秒为单位

例如，实现一个持续 1s 以物件中点为中心顺时针旋转 180°效果的动画，代码如下：

```
<?xml version="1.0" encoding="utf-8"?>
<set xmlns:android="http://schemas.android.com/apk/res/android" >
    <rotate
        android:duration="1000"
        android:fromDegrees="0"
        android:interpolator="@android:anim/accelerate_decelerate_
                             interpolator"
        android:pivotX="50%"
        android:pivotY="50%"
        android:toDegrees="+180" />
</set>
```

③ scale 渐变尺寸伸缩动画效果属性如表 9.9 所示。

表 9.9 scale 属性描述

属性	说明	参数说明
fromXScale	动画起始时 X 坐标上的伸缩尺寸	0.0 表示收缩到没有
fromYScale	动画起始时 Y 坐标上的伸缩尺寸	1.0 表示正常无伸缩
toXScale	动画结束时 X 坐标上的伸缩尺寸	值小于 1.0 表示收缩
toYScale	动画结束时 Y 坐标上的伸缩尺寸	值大于 1.0 表示放大
pivotX	动画相对于物件的 X 坐标的开始位置	从 0%～100%中取值
pivotY	动画相对于物件的 Y 坐标的开始位置	50%为物件的 X 或 Y 方向坐标上的中点位置
duration	动画持续时间	时间以毫秒为单位
fillAfter	当设置为 true 时，该动画转化在动画结束后被应用	

例如，实现一个持续 1s 由正常缩放到 50%大小效果的动画，代码如下：

```
<?xml version="1.0" encoding="utf-8"?>
<set xmlns:android="http://schemas.android.com/apk/res/android" >
```

```
    <scale
        android:duration="1000"
        android:fillAfter="false"
        android:fromXScale="1.0"
        android:fromYScale="1.0"
        android:pivotX="50%"
        android:pivotY="50%"
        android:toXScale="0.5"
        android:toYScale="0.5" />
</set>
```

④ translate 画面转换位置移动动画效果的属性如表 9.10 所示。

表 9.10 translate 属性描述

属性	说明	参数说明
fromXDelta	动画起始时 X 坐标上的位置	没有指定的时候，默认是以自己为相对参照物
fromYDelta	动画起始时 Y 坐标上的位置	
toXDelta	动画结束时 X 坐标上的位置	
toYDelta	动画结束时 Y 坐标上的位置	
duration	动画持续时间	时间以毫秒为单位

例如，实现一个持续 2s 移动效果的动画，代码如下：

```
<?xml version="1.0" encoding="utf-8"?>
<set xmlns:android="http://schemas.android.com/apk/res/android" >
    <translate
        android:duration="2000"
        android:fromXDelta="0"
        android:fromYDelta="0"
        android:toXDelta="180"
        android:toYDelta="180" />
</set>
```

XML 文件中必须至少有一个根元素，可以是<alpha>、<rotate>、<scale>、<translate>中的任意一个，也可以是<set>管理一个由 4 个根元素中任意几个根元素组成的动画集合。<set>是可以嵌套的。

有人曾形象地把<set>比喻为一个动画容器，当存在多个动画元素时，默认这些动画同时发生，但也可通过 startOffset 属性设置各个动画的开始时间，达到多个动画元素顺序播放的效果。

以上的方法可以决定动画的开始状态、结束状态以及变化效果，通过设置的动画持续时间，就可以算出在中间插入多少帧的动画实现动画效果。但是，在动画运行时，动画效果的速率还需要借助于 Interpolator 来控制。Interpolator 可译为"插值器"，用于控制插入帧的属性，它依据特定的算法可以计算出整个动画所需要插入帧的密度与位置，从而使基本的动画实现匀速、加速、减速等各种速度变化的效果。Interpolator 接口的常用实现类如表 9.11 所示。

表 9.11　Interpolator 接口的常用实现类

方法	简述
AccelerateDecelerateInterpolator	动画开始、结束时减速，中间加速
AccelerateInterpolator	动画开始时，速度较慢，然后开始加速
CycleInterpolator	动画循环播放指定次数，速度按正弦曲线改变
DecelerateInterpolator	动画开始时，速度较快，然后开始减速
LinearInterpolator	动画以匀速变化

为了在动画资源中指定补间动画所使用的插值器，可以定义补间动画的<set/>元素支持 android:interpolator 属性，设置相应的参数就可以控制动画效果的速率。android 提供的三种动画插值器如表 9.12 所示。

表 9.12　Interpolator 接口的常用实现类

属性	简述
accelerate_decelerate_interpolator	加速-减速动画插值器
accelerate_interpolator	加速动画插值器
decelerate_interpolator	减速动画插值器

【例 9.4】　实现补间动画实例

下面运用以上提供的各种动画属性实现一个动画切换的程序。

定义 Activity 程序：

```java
package com.example.tweenanimation;
import android.support.v7.app.AppCompatActivity;
import android.os.Bundle;
import android.view.View;
import android.view.animation.Animation;
import android.view.animation.AnimationUtils;
import android.widget.Button;
import android.widget.ImageView;
public class MainActivity extends AppCompatActivity {
    private Button btnAlpha,btnScale,btnTrans,btnRotate,btnAll;
    private ImageView mImg;
    private Animation mAnimation;
    @Override
    protected void onCreate(Bundle savedInstanceState) {
        super.onCreate(savedInstanceState);
        setContentView(R.layout.activity_main);
        btnAll = (Button) findViewById(R.id.btnAll);
        btnAlpha = (Button) findViewById(R.id.btnAlpha);
        btnScale = (Button) findViewById(R.id.btnScale);
        btnRotate = (Button) findViewById(R.id.btnRotate);
        btnTrans = (Button) findViewById(R.id.btnTrans);
        mImg = (ImageView) findViewById(R.id.imageView1);
        btnAlpha.setOnClickListener(new View.OnClickListener() {
            @Override
            public void onClick(View v) {
```

```java
            mAnimation = AnimationUtils.loadAnimation(MainActivity.this
                    , R.anim.my_alpha);
            mImg.startAnimation(mAnimation);
        }
    });
    btnScale.setOnClickListener(new View.OnClickListener() {
        @Override
        public void onClick(View v) {
            mAnimation = AnimationUtils.loadAnimation(MainActivity.this
                    , R.anim.my_scale);
            mImg.startAnimation(mAnimation);
        }
    });
    btnRotate.setOnClickListener(new View.OnClickListener() {
        @Override
        public void onClick(View v) {
            mAnimation = AnimationUtils.loadAnimation(MainActivity.this
                    , R.anim.my_rotate);
            mImg.startAnimation(mAnimation);
        }
    });
    btnTrans.setOnClickListener(new View.OnClickListener() {
        @Override
        public void onClick(View v) {
            mAnimation = AnimationUtils.loadAnimation(MainActivity.this
                    , R.anim.my_translate);
            mImg.startAnimation(mAnimation);
        }
    });
    btnAll.setOnClickListener(new View.OnClickListener() {
        @Override
        public void onClick(View v) {
            mAnimation = AnimationUtils.loadAnimation(MainActivity.this
                    , R.anim.my_all);
            mImg.startAnimation(mAnimation);
        }
    });
  }
}
```

其中的<alpha>、<rotate>、<scale>、<translate>在之前有所展示,这里不再重复。<set>同时包含多个动画的元素的方法如下:

```xml
<?xml version="1.0" encoding="utf-8"?>
<set
    xmlns:android="http://schemas.android.com/apk/res/android" >
    <alpha
        android:duration="1000"
```

第9章 图形图像与多媒体

```
            android:fromAlpha="0.5"
            android:toAlpha="1.0" />
        <rotate
            android:duration="1000"
            android:fromDegrees="0"
            android:pivotX="50%"
            android:pivotY="50%"
            android:toDegrees="+180" />
        <scale
            android:duration="1000"
            android:fillAfter="false"
            android:fromXScale="1.0"
            android:fromYScale="1.0"
            android:pivotX="50%"
            android:pivotY="50%"
            android:toXScale="0.5"
            android:toYScale="0.5" />
        <translate
            android:duration="1000"
            android:fromXDelta="0"
            android:fromYDelta="0"
            android:toXDelta="180"
            android:toYDelta="180"/>
</set>
```

定义布局文件：

```
<LinearLayout xmlns:android="http://schemas.android.com/apk/res/android"
    xmlns:tools="http://schemas.android.com/tools"
    android:layout_width="fill_parent"
    android:layout_height="fill_parent"
    android:orientation="vertical" >
    <ImageView
        android:id="@+id/imageView1"
        android:layout_width="200dp"
        android:layout_height="200dp"
        android:layout_gravity="center_horizontal"
        android:src="@drawable/pic" />
    <LinearLayout
        android:layout_width="fill_parent"
        android:layout_height="wrap_content"
        >
        <Button
            android:id="@+id/btnAlpha"
            android:layout_width="wrap_content"
            android:layout_height="wrap_content"
            android:text="Alpha" />
        <Button
```

```xml
            android:id="@+id/btnRotate"
            android:layout_width="wrap_content"
            android:layout_height="wrap_content"
            android:text="Rotate" />
        <Button
            android:id="@+id/btnScale"
            android:layout_width="wrap_content"
            android:layout_height="wrap_content"
            android:text="Scale" />
        <Button
            android:id="@+id/btnTrans"
            android:layout_width="wrap_content"
            android:layout_height="wrap_content"
            android:text="Translate" />
    </LinearLayout>
    <Button
        android:id="@+id/btnAll"
        android:layout_width="wrap_content"
        android:layout_height="wrap_content"
        android:text="All" />
</LinearLayout>
```

效果如图 9.5 所示。

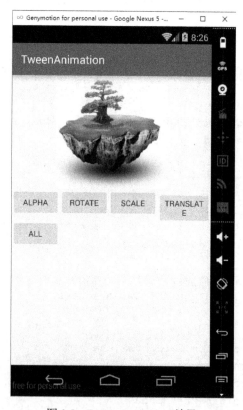

图 9.5　Frame Animation 效果

9.3.3 Property Animation

Property Animation，即属性动画。与 Tween Animation 不同的是，前者更改对象的实际属性，后者改变 View 的绘制效果，而不改变 View 的属性。

在 Property Animation 中，有如表 9.13 所示的动画操作属性。

表 9.13 Property Animation 动画操作属性

属性	简述
Duration	动画持续时间，默认 300ms
TimeInterpolation	定义动画的变化速率
Repeat count and behavior	定义重复次数、重复模式
Animator sets	动画集合
Frame refresh delay	帧刷新延迟

9.3.4 动画监听器：AnimationListener

AnimationListener，即动画监听器，接收动画发出的表明其相关事件的通知，如动画结束或重复播放等。在 Android 系统中，有一个特定的接口用来监听动画的操作状态：android.view.animation.Animation.AnimationListener，其定义方法如表 9.14 所示。

表 9.14 AnimationListener 接口定义方法

返 回 类 型	方法	简述
abstract void	onAnimationCancel（Animator animation）	动画取消播放时被调用
	onAnimationEnd（Animator animation）	动画结束时被调用
	onAnimationRepeat（Animator animation）	动画重复播放时被调用
	onAnimationStart（Animator animation）	动画开始时被调用

9.4 在 Android 中播放音频与视频

智能手机能够播放音乐、视频文件已经成为不可或缺的多媒体功能。Android 在其多媒体框架中支持多种常见的媒体类型，所以开发者能够轻松地将音频、视频以及图像集成到应用程序中，用户能通过多种形式播放音视频文件，可以利用 MediaPlayer 类提供的 API 直接读取应用中的资源文件（raw 文件）、文件系统中的单独文件，也可以从互联网上获取数据流。

9.4.1 MediaPlayer 介绍

在介绍 MediaPlayer 类之前先介绍 MediaPlayer 所支持操作的生命周期，图 9.6 展示了其生命周期，这张图清晰地描述了 MediaPlayer 的各个状态，也列举了主要方法的调用时序，每种方法只能在一些特定的状态下使用。

（1）Idle 状态：当使用 new() 方法创建了一个 MediaPlayer 对象或者调用了其 reset() 方法时，该 MediaPlayer 对象处于 idle 状态。

（2）End 状态：通过 release() 方法可以进入 End 状态，只要 MediaPlayer 对象不再被使用，就应当尽快将其通过 release() 方法释放掉，以释放相关的软硬件资源，进入 End 状态后将不会进入任何其他的状态了。

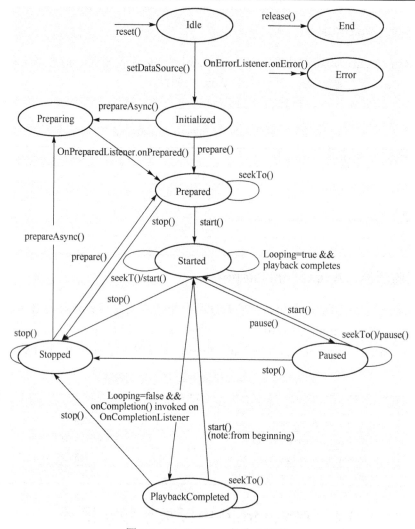

图 9.6 MediaPlayer 的生命周期

（3）Initialized 状态：当 MediaPlayer 调用 setDataSource()方法进入 Initialized 状态时，表示此时要播放的文件已经设置好了。

（4）Prepared 状态：初始化之后还需要调用 prepare()方法(同步)或 prepareAsync()方法(异步)进入此状态，即预播放状态表示目前的媒体文件没有任何问题可以进行文件播放。

（5）Started 状态：MediaPlayer 一旦准备好，就可以调用 start()方法，文件进入播放状态，可以使用 isPlaying()方法检测 MediaPlayer 是否处于 Started 状态。在该状态下，可以使用 seekTo()方法制定媒体播放位置。

（6）Paused 状态：在 Started 状态下或者调用 pause()方法可以暂停媒体播放，从而进入 Paused 状态，再使用 start()方法可以继续媒体播放，暂停时仍然可以使用 seekTo()方法。

（7）Stopped 状态：在 Started 状态或者 Paused 状态下可调用 stop()停止媒体播放，而想要重新播放处于停止状态的媒体文件，需要通过 prepareAsync()和 prepare()方法进入 Prepared 状态。

（8）PlaybackCompleted 状态：当媒体文件正常播放完毕，并且没有设置循环播放之后就会进入此状态，并会触发 OnCompletionListener 的 onCompletion() 方法，可以通过 stop() 方法停止播放，也可以通过 start() 方法重新播放，或者使用 seekTo() 方法来重新定位播放位置。

（9）Error 状态：当用户播放操作中出现错误时会触发 OnErrorListener.onError() 事件，此时媒体播放进入 Error 状态，可以通过 reset() 方法重新返回 Idle 状态。

MediaPlayer 类的常用方法表 9.15 所示。

表 9.15　MediaPlayer 类的常用方法

返回类型	方法	简述
boolean	isLooping()	检查 MediaPlayer 是否循环播放
	isPlaying()	检查 MediaPlayer 是否正在播放
int	getAudioSessionId()	返回音频的会话 ID
	getCurrentPosition()	取得当前播放的位置点
	getDuration()	取得文件的长度
	getVideoHeight()	返回视频的高度
	getVideoWidth()	返回视频的宽度
static MediaPlayer	create(Context context, int resid)	从指定的源 id 中创建一个 MediaPlayer 对象
	create(Context context, Uri uri)	从指定的 Uri 中创建一个 MediaPlayer 对象
	create(Context context, Uri uri, SurfaceHolder holder)	从指定的 Uri 中创建一个 MediaPlayer 对象，并在 Surfaceview 中显示
void	pause()	暂停播放
	prepare()	播放前同步准备播放器
	prepareAsync()	播放前异步准备播放器
	release()	释放此 MediaPlayer 对象占用的资源
	reset()	重置 MediaPlayer 到未初始化状态
	seekTo(int msec)	寻找特定的播放点
	selectTrack(int index)	选择路径
	setAudioAttributes(AudioAttributes attributes)	设置此 MediaPlayer 的音频属性
	setAudioSessionId(int sessionId)	设置音频的会话 ID
	setAudioStreamType(int streamtype)	设置音频类型
	setDataSource(String path)	设置数据文件(file-path 或 http/rtsp URL)
	setDataSource(Context context, Uri uri)	指定数据源
	setDisplay(SurfaceHolder sh)	设置视频显示
	setLooping(boolean looping)	设置循环
	setOnCompletionListener(MediaPlayer.OnCompletionListener listener)	媒体播放完成后触发
	setOnErrorListener(MediaPlayer.OnErrorListener listener)	出现错误时被触发
	setOnPreparedListener(MediaPlayer.OnPreparedListener listener)	当媒体准备完成后触发
	setOnSeekCompleteListener(MediaPlayer.OnSeekCompleteListener listener)	当媒体设置播放点后触发
	setOnVideoSizeChangedListener(MediaPlayer.OnVideoSizeChangedListener listener)	当视频文件大小改变后触发
	setVideoScalingMode(int mode)	设置视频缩放比例
	setVolume(float leftVolume, float rightVolume)	设置播放音量
	start()	开始或恢复播放
	stop()	停止播放

9.4.2 运用 MediaPlayer 播放音频

在熟悉了 MediaPlayer 类的基本方法之后，下面展示如何运用 MediaPlayer 播放一个 MP3 音频文件。本例中使用了"播放"、"暂停"、"停止"三个按钮来控制音频播放且使用了一个 seekBar 来表示播放的进度，并且可以拖动进度条设置播放位置。需要播放的音频文件为 my_music.mp3，该文件保存在 res\raw\文件夹下。

【例 9.5】 MediaPlayer 播放 MP3 实例

定义布局文件：

```xml
<?xml version="1.0" encoding="utf-8"?>
<LinearLayout xmlns:android="http://schemas.android.com/apk/res/android"
    xmlns:tools="http://schemas.android.com/tools"
    android:layout_width="match_parent"
    android:layout_height="match_parent"
    android:orientation="vertical">
    <LinearLayout
        android:layout_width="wrap_content"
        android:layout_height="wrap_content"
        android:orientation="horizontal">
        <TextView
            android:layout_width="wrap_content"
            android:layout_height="wrap_content"
            android:text="@string/play_state"
            android:textSize="25sp"
            />
        <TextView
            android:id="@+id/tvState"
            android:layout_width="wrap_content"
            android:layout_height="wrap_content"
            android:text="@string/stete_prepare"
            android:textSize="25sp"
            />
    </LinearLayout>

    <LinearLayout
        android:layout_width="match_parent"
        android:layout_height="wrap_content"
        android:layout_marginTop="10dp"
        android:orientation="horizontal">
        <Button
            android:id="@+id/btnStart"
            android:layout_width="wrap_content"
            android:layout_height="wrap_content"
            android:layout_marginLeft="5dp"
            android:text="@string/start"/>
        <Button
```

```xml
            android:id="@+id/btnPause"
            android:layout_width="wrap_content"
            android:layout_height="wrap_content"
            android:layout_marginLeft="5dp"
            android:text="@string/puase"/>
        <Button
            android:id="@+id/btnEnd"
            android:layout_width="wrap_content"
            android:layout_height="wrap_content"
            android:layout_marginLeft="5dp"
            android:text="@string/end"/>
        <Button
            android:id="@+id/btnRestart"
            android:layout_width="wrap_content"
            android:layout_height="wrap_content"
            android:layout_marginLeft="5dp"
            android:text="@string/restart"/>
    </LinearLayout>
    <SeekBar
        android:layout_width="match_parent"
        android:layout_height="wrap_content"
        android:layout_marginTop="10dp"
        android:id="@+id/seekBar"
        android:layout_gravity="center_horizontal"
        android:indeterminate="false" />
</LinearLayout>
```

然后定义 Activity 程序,在 Activity 程序中,主要是对其中的几个按钮以及 SeekBar 定义监听,其次是对于 SeekBar 的控制,运用一个异步线程刷新 MP3 的播放情况。

```java
package com.example.mediaplayer;
import android.app.Activity;
import android.media.MediaPlayer;
import android.os.AsyncTask;
import android.os.Bundle;
import android.view.View;
import android.view.View.OnClickListener;
import android.widget.Button;
import android.widget.SeekBar;
import android.widget.TextView;
public class MainActivity extends Activity implements OnClickListener{
    private MediaPlayer mediaPlayer;
    private TextView tvState;
    private Button btnPlay,btnPause,btnStop,btnRestart;
    private SeekBar seekBar;
    private boolean playFlag = false,pauseFlag = false;
    @Override
    protected void onCreate(Bundle savedInstanceState) {
```

```java
        super.onCreate(savedInstanceState);
        setContentView(R.layout.activity_main);
        btnPlay = (Button)this.findViewById(R.id.btnStart);
        btnPause = (Button) this.findViewById(R.id.btnPause);
        btnStop = (Button) this.findViewById(R.id.btnEnd);
        btnRestart= (Button) this.findViewById(R.id.btnRestart);
        tvState= (TextView)this.findViewById(R.id.tvState);
        seekBar = (SeekBar)this.findViewById(R.id.seekBar);
        btnPlay.setOnClickListener(this);
        btnPause.setOnClickListener(this);
        btnStop.setOnClickListener(this);
        btnRestart.setOnClickListener(this);
        mediaPlayer = MediaPlayer.create(MainActivity.this, R.raw.my_music);
    }

/**
*在按键监听函数中运用start()、stop()、pause()、release()方法实现对播放的控
*制
*/
    @Override
    public void onClick(View v) {
    switch (v.getId())
    {
        case R.id.btnStart:
            if(pauseFlag==false) {
                mediaPlayer.setOnCompletionListener(new MediaPlayer
                .OnCompletionListener() {
                    @Override
                    public void onCompletion(MediaPlayer mp) {
                        tvState.setText(getResources().getString(R.string.stete_end));
                        playFlag = false;
                        mp.release();
                        }
                });
                seekBar.setMax(mediaPlayer.getDuration());
                playFlag = true;
                UpdateSeekBar updateProgress = new UpdateSeekBar();
                updateProgress.execute(1000);
                seekBar.setOnSeekBarChangeListener(new
                    MySeekBarChangeListener());
                try {
                    if (mediaPlayer != null) {
                        mediaPlayer.stop();
                    }
                        mediaPlayer.prepare();
                        mediaPlayer.start();
                        tvState.setText(getResources().getString(R.string.start));
```

```java
            } catch (Exception e) {
            this.tvState.setText(getResources()
                .getString(R.string.stete_error));
            }
        }else {
            pauseFlag = false;
            playFlag = true;
            mediaPlayer.start();
            UpdateSeekBar updateProgress = new UpdateSeekBar();
            updateProgress.execute(1000);
            seekBar.setOnSeekBarChangeListener(new
                MySeekBarChangeListener());
            tvState.setText(getResources().getString(R.string.start));
        }
        break;
    case R.id.btnEnd:
        if(mediaPlayer.isPlaying()){
            mediaPlayer.stop();
            playFlag = false;
            tvState.setText(getResources().getString(R.string.end));
        }
        break;
    case R.id.btnPause:
        if(mediaPlayer.isPlaying())
        {
            playFlag = false;
            pauseFlag=true;
            mediaPlayer.pause();
            tvState.setText(getResources().getString(R.string.puase));
        }
        break;
    case R.id.btnRestart:
        if(mediaPlayer.isPlaying()){
            playFlag = true;
            mediaPlayer.seekTo(0);
        }
        break;
    }
}
/**
 *在异步线程中实现对音乐播放进度的显示
 */
private class UpdateSeekBar extends AsyncTask<Integer,Integer,String> {
    @Override
    protected void onPostExecute(String s) {
    }
    @Override
```

```java
        protected void onProgressUpdate(Integer... values) {
            seekBar.setProgress(values[0]);
        }
        @Override
        protected String doInBackground(Integer... params) {
            while(playFlag){
                try {
                    Thread.sleep(params[0]);
                } catch (InterruptedException e) {
                    e.printStackTrace();
                }
                this.publishProgress(mediaPlayer.getCurrentPosition());
            }
            return null;
        }
    }
    /**
     *在SeekBar监听中实现对音乐播放的控制
     */
    private class MySeekBarChangeListener implements SeekBar.
            OnSeekBarChangeListener {
        @Override
        public void onProgressChanged(SeekBar seekBar, int progress,
                boolean fromUser) {
        }
        @Override
        public void onStartTrackingTouch(SeekBar seekBar) {
        }
        @Override
        public void onStopTrackingTouch(SeekBar seekBar) {
            mediaPlayer.seekTo(seekBar.getProgress());
        }
    }
}
```

在Genymotion模拟器中，运行效果如图9.7所示。

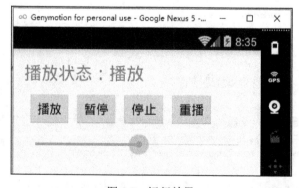

图9.7　运行效果

9.4.3 播放视频

在 Android 中播放视频有三种方式。
- 使用 Android 自带的播放器。
- 使用 VideoView 来播放视频。
- 使用 MediaPlayer 类和 SurfaceView 来实现。

下面介绍使用以上三种方式播放一段视频的方法，文件放置在 SD 卡根目录下的 Download 文件目录下，本实例中向模拟器中导入视频的方式有以下两种。

方式一：需要打开 Android Device Monitor，向模拟器的文件中导入视频文件，如图 9.8 所示打开 Android Device Monitor。

图 9.8　打开 Android Device Monitor

选中 "File Explorer" 单击右方的 "push a file onto the device" 选择相关的文件，就能将其导入到虚拟机的/sdcard/Download 文件目录下。

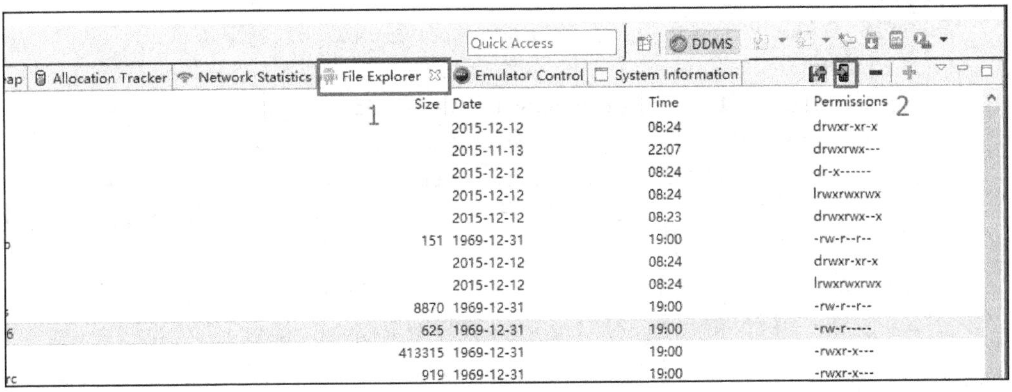

图 9.9　导入文件

方式二：打开虚拟机后可直接将文件拖到虚拟机中文件，即到方式一中相同的文件目录下。导入后的文件如图 9.10 所示。

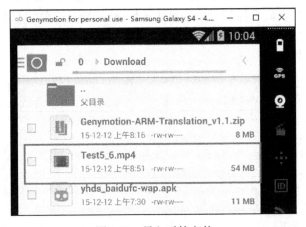

图 9.10　导入后的文件

注意：在调用 SD 卡文件时需要声明权限，不然会出现无法播放的情况。在工程的 Manifest.xml 文件中加入以下代码，实现对 SD 卡的读/写操作。

```
    <uses-permission android:name=
"android.permission.WRITE_EXTERNAL_STORAGE"/>
    <uses-permission android:name=
"android.permission.MOUNT_UNMOUNT_FILESYSTEMS"/>
```

（1）使用 Android 自带播放器播放，指定 Action 为 ACTION_VIEW，Data 为 Uri。

【例 9.6】 Android 自带播放器播放视频实例

在 Activity 程序中定义如下代码。

```java
package com.example.videoplayer;
import android.content.Intent;
import android.net.Uri;
import android.os.Environment;
import android.support.v7.app.AppCompatActivity;
import android.os.Bundle;
public class MainActivity extends AppCompatActivity {
    @Override
    protected void onCreate(Bundle savedInstanceState) {
        super.onCreate(savedInstanceState);
        setContentView(R.layout.activity_main);
        Uri uri = Uri.parse(Environment.getExternalStorageDirectory()
            .getPath()+"/Download/Test5_6.mp4");
        Intent it = new Intent(Intent.ACTION_VIEW);
        it.setDataAndType(uri,"video/mp4");
        startActivity(it);
    }
```

执行程序之后，程序立即打开一个跳转页面选择播放器。

（2）使用 VideoView 播放视频，这个控件使用起来十分方便，首先在布局文件中创建该组件，然后在 Activity 中获取该组件，再通过 setVideoURI()方法或者 setVideoPath()方法加载要播放的视频文件，最后类似于播放音乐的模式使用 start()方法播放视频。

【例 9.7】 VideoView 实现视频播放实例

首先定义布局文件：

```xml
<?xml version="1.0" encoding="utf-8"?>
<LinearLayout xmlns:android="http://schemas.android.com/apk/res/android"
    xmlns:tools="http://schemas.android.com/tools"
    android:layout_width="match_parent"
    android:layout_height="match_parent"
    android:orientation="vertical"
    tools:context="com.example.videoview.MainActivity"
    android:weightSum="1">
    <VideoView
        android:layout_width="match_parent"
```

```
            android:layout_height="match_parent"
            android:id="@+id/videoView" />
</LinearLayout>
```

然后定义 Activity 程序：

```
package com.example.videoview;
import android.net.Uri;
import android.os.Environment;
import android.support.v7.app.AppCompatActivity;
import android.os.Bundle;
import android.widget.MediaController;
import android.widget.VideoView;
public class MainActivity extends AppCompatActivity {
    VideoView myVideo;
    @Override
    protected void onCreate(Bundle savedInstanceState) {
        super.onCreate(savedInstanceState);
        setContentView(R.layout.activity_main);
        MediaController myMc = new MediaController(MainActivity.this);
        Uri uri = Uri.parse(Environment.getExternalStorageDirectory()
            .getPath()+"/Download/Test5_6.mp4");
        myVideo = (VideoView)this.findViewById(R.id.videoView);
        myVideo.setVideoURI(uri);
        myVideo.setMediaController(myMc);
        myVideo.requestFocus();
        try{
            myVideo.start();
        }catch (Exception e){
            e.printStackTrace();
        }
    }
}
```

尽管 VideoView 在使用的过程中十分方便，但是由于 VideoView 在视频格式上有一定条件的限制，只有少数满足其格式要求的视频文件才能被完美地播放，所以在实际运用中不推荐使用这种方式。

(3) 使用 MediaPlayer 和 SurfaceView 播放视频。

在上一节中展示了如何运用 MediaPlayer 播放音乐文件，同样的 MediaPlayer 也被用于播放视频文件，但是仅靠 MediaPlayer 是不够的，因为它没有提供视频播放的界面。这时，就需要用 SurfaceView 创建视频播放的界面，SurfaceView 组件能够提供快速的 GUI 更新，并且启动一个新的线程处理视频播放中的图形加载，使得视频播放更为流畅。如果要设置播放界面的大小，则还需要 SurfaceHolder 接口提供某些方法。表 9.16 展示了相关的方法。

表 9.16　SurfaceHolder 类定义的方法

返回类型	方法	简述
abstract Surface	getSurface()	界面对象的直接接口

返回类型	方法	简述
abstract void	setFixedSize (int width, int height)	设置界面的固定尺寸
	setKeepScreenOn (boolean screenOn)	设置界面显示时屏幕打开
	setSizeFromLayout ()	允许界面基于其容器调整大小
	setType (int type)	设置界面的类型,一般自动设置

【例9.8】 SurfaceView 与 MediaPlayer 实现视频播放实例

下面展示如何运用 SurfaceView 与 MediaPlayer 完成视频播放。

首先建立布局文件:

```xml
<?xml version="1.0" encoding="utf-8"?>
<LinearLayout xmlns:android="http://schemas.android.com/apk/res/android"
    xmlns:tools="http://schemas.android.com/tools"
    android:layout_width="match_parent"
    android:layout_height="match_parent"
    android:paddingBottom="@dimen/activity_vertical_margin"
    android:paddingLeft="@dimen/activity_horizontal_margin"
    android:paddingRight="@dimen/activity_horizontal_margin"
    android:paddingTop="@dimen/activity_vertical_margin"
    android:orientation="vertical"
    tools:context="com.example.surfaceview.MainActivity" >
    <SurfaceView
        android:layout_width="match_parent"
        android:layout_height="200dp"
        android:id="@+id/mySurface"/>
    <Button
        android:layout_width="wrap_content"
        android:layout_height="wrap_content"
        android:text="@string/tv_play"
        android:id="@+id/btnStart"/>
    <Button
        android:layout_width="wrap_content"
        android:layout_height="wrap_content"
        android:text="@string/tv_pause"
        android:id="@+id/btnPause"/>
    <Button
        android:layout_width="wrap_content"
        android:layout_height="wrap_content"
        android:text="@string/tv_stop"
        android:id="@+id/btnStop"/>
</LinearLayout>
```

在 Activity 程序中,让其继承于 SurfaceHolder.Callback,使得在创建视频时复写其方法实现自动加载。再通过"播放"、"暂停"、"停止"三个按钮控制视频的播放。具体实现如下:

```java
package com.example.surfaceview;
import android.app.Activity;
import android.media.AudioManager;
import android.media.MediaPlayer;
import android.net.Uri;
```

```java
import android.os.Environment;
import android.os.Bundle;
import android.view.SurfaceHolder;
import android.view.SurfaceView;
import android.view.View;
import android.widget.Button;
public class MainActivity extends Activity implements SurfaceHolder.Callback{
    MediaPlayer mp;
    SurfaceView mySurface;
    SurfaceHolder surfaceHolder;
    Button btnPlay,btnPause,btnStop;
    @Override
    public void onCreate(Bundle savedInstanceState) {
        super.onCreate(savedInstanceState);
        setContentView(R.layout.activity_main);
        btnPlay=(Button)findViewById(R.id.btnStart);
        btnPause=(Button)findViewById(R.id.btnPause);
        btnStop=(Button)findViewById(R.id.btnStop);
        mySurface=(SurfaceView)findViewById(R.id.mySurface);
        mp = new MediaPlayer();
        surfaceHolder=mySurface.getHolder();
                //SurfaceHolder是SurfaceView的控制接口
        surfaceHolder.setKeepScreenOn(true);
        surfaceHolder.setFixedSize(320,220);
        surfaceHolder.setType(SurfaceHolder
                .SURFACE_TYPE_PUSH_BUFFERS);    //Surface类型
        surfaceHolder.addCallback(this);
        btnPlay.setOnClickListener(new View.OnClickListener() {
            @Override
            public void onClick(View v) {
                play();
            }});
        btnPause.setOnClickListener(new View.OnClickListener(){
            @Override
            public void onClick(View v) {
                mp.pause();
            }});
        btnStop.setOnClickListener(new View.OnClickListener(){
            @Override
            public void onClick(View v) {
                mp.stop();
            }});
    }
    @Override
    protected void onDestroy() {
        //TODO Auto-generated method stub
        super.onDestroy();
        if(mp.isPlaying()){
            mp.stop();
        }       //Activity销毁时停止播放，释放资源,
                    不做这个操作，即使退出仍能听到视频播放的声音
```

```
            mp.release();
        }
        public void play(){
            try {
                mp.reset();
                mp.setAudioStreamType(AudioManager.STREAM_MUSIC);
                Uri uri = Uri.parse(Environment.getExternalStorageDirectory()
                        .getPath()+"/Test1.mp4");
                mp.setDataSource(MainActivity.this,uri);
                mp.setDisplay(mySurface.getHolder());
                mp.prepare();
                mp.start();
            }
            catch (Exception e){
                e.printStackTrace();
            }
        }
        @Override
        public void surfaceCreated(SurfaceHolder holder) {
            try {
                play();
            } catch (Exception e) {
                //TODO Auto-generated catch block
                e.printStackTrace();
            }
        }

        @Override
    public void surfaceChanged(SurfaceHolder holder, int format,
            int width, int height) {
        }

        @Override
        public void surfaceDestroyed(SurfaceHolder holder) {
        }
    }
```

视频播放效果如图 9.11 所示。

图 9.11 视频播放效果

9.5 控制相机拍照

Camera 类可用来设置图像捕捉的相关参数、启动/停止预览、拍照和检索视频的帧的编码。相当于一个 Camera 服务的客户端,管理实际的摄像头硬件。

在访问设备的相机之前,必须在 Android Manifest 中声明 CAMERA 权限。当然,也要包括<uses-feature>manifest 元素声明的摄像头功能由用户的应用程序使用。比如,使用摄像头和自动对焦功能,Manifest 应包括如下内容:

```
<uses-permission android:name="android.permission.CAMERA" />
<uses-feature android:name="android.hardware.camera" />
<uses-feature android:name="android.hardware.camera.autofocus" />
```

使用 Camera 类控制摄像头拍照的步骤如下。

(1) 从 open(int)中获取一个摄像头的实例。
(2) 在 getparameters()中得到默认设置参数。
(3) 如果有必要,可修改返回的 Camera.Parameters 对象,并调用

```
setParameters(Camera.Parameters)
```

(4) 如果需要,可调用 setdisplayorientation(int)。
(5) 将一个完全初始化的 SurfaceHolder 传递给 setPreviewDisplay(SurfaceHolder)十分重要。否则,相机将无法启动预览。
(6) 调用 startpreview()开始更新预览表。预览开始之前,必须拍摄一张照片。
(7) 可以调用 takePicture(Camera.ShutterCallback, Camera.PictureCallback, Camera.PictureCallback, Camera.PictureCallback)捕捉照片。等待回调函数提供实际的图像数据。
(8) 拍照后,预览显示将停止。如需拍摄更多照片,可再次调用 startPreview()。
(9) 调用 stoppreview()来停止更新预览表。
(10) 调用 release()释放由其他应用程序使用的摄像头。应用程序应该在 onpause()中立即释放摄像头。

Camera 类的常用方法如表 9.17 所示。

表 9.17 Camera 类的常用方法

返回类型	方法	简述
Camera.Parameters	getParameters()	返回摄像头的参数
final void	autoFocus(Camera.AutoFocusCallback cb)	自动对焦,且当聚焦后记录一个回调函数
	cancelAutoFocus()	取消进程中的所有自动对焦
	lock()	锁定摄像头,阻止其他程序访问
	reconnect()	在其他程序调用摄像头后,重新连接摄像头服务
	release()	断开连接并释放摄像头资源
	startPreview()	开始捕捉、预览
	stopPreview()	停止捕捉、预览
	setDisplayOrientation(int degrees)	设置摄像头角度
	setZoomChangeListener(Camera.OnZoomChangeListener listener)	显示区域发生改变时触发

返回类型	方法	简述
final void	takePicture (Camera.ShutterCallback shutter, Camera.PictureCallback raw, Camera.PictureCallback jpeg)	捕捉图像
	unlock ()	解锁摄像头，允许其他程序访问
static int	getNumberOfCameras ()	返回物理摄像头的个数
static void	getCameraInfo (int cameraId, Camera.CameraInfo cameraInfo)	返回指定摄像头的信息
static Camera	open (int cameraId)	创建新的 Camera 对象以打开指定的摄像头
	open ()	创建新的 Camera 对象以打开默认的摄像头
void	setParameters (Camera.Parameters params)	设置摄像头的参数

SurfaceHolder.Callback 接口定义的方法如表 9.18 所示。

表 9.18　SurfaceHolder.Callback 接口定义的方法

返回类型	方法	简述
abstract void	surfaceChanged (SurfaceHolder holder, int format, int width, int height)	当界面的格式和大小发生改变时被调用
	surfaceCreated (SurfaceHolder holder)	当界面创建后被调用
	surfaceDestroyed (SurfaceHolder holder)	当界面关闭前被调用

Camera 类定义的接口如表 9.19 所示。

表 9.19　Camera 类定义的接口

方法	简述
Camera.AutoFocusCallback	自动对焦的回调接口
Camera.ErrorCallback	出现错误时的回调接口
Camera.FaceDetectionListener	预览时检测到人脸的回调接口
Camera.OnZoomChangeListener	显示区域改变时的回调接口
Camera.PictureCallback	图片捕捉后提供数据的回调接口
Camera.PreviewCallback	预览时的回调接口
Camera.ShutterCallback	标记实际捕捉图像操作的回调接口
SensorEventListener	传感器参数改变时的回调接口

第 10 章 Android 的网络编程基础

在 PC 时代，以太网为人们提供了极大的便利，各种信息服务使得计算机成为不可或缺的工具。随着移动终端的普及，移动互联网带来的便捷性也不言而喻，网上购物、网络游戏、即时通信等都需要网络技术的支持。在这一章中，将由浅及深地讲解 Android 的网络编程方法，从最基础的 TCP/IP 网络通信协议实现数据交流，到使用 WebService 技术访问网络数据，并开发一款手机归属地查询小应用。

10.1 基于 TCP 协议的网络通信

在网络环境中，由程序创建的"字节序列"被称为"分组报文"。一组报文包括网络用来完成工作的控制信息，还包括数据内容。协议相当于互联网中相互通信的程序间的一种约定，它规定了分组报文数据交换的方式与含义。

在网络层中，IP 协议完成将分组报文传输到指定的目的地的任务，这时网络通信完成了点对点（Prot to Port）的传输，但并不能精确到程序。在传输层中，TCP 协议负责更细的寻址操作，将报文发送到对应的程序中，这就完成了端对端（End to End）的传输。

在网络连接中最基础的两个协议就是 TCP/IP 与 UDP，前者是面向连接的协议，能够为数据提供高效可靠的传输机制；而后者则是面向无连接的广播通信机制，适用于数据量少、对通信质量不高的环境。在 Android 网络编程中，为保证数据通信安全可靠，大都使用 TCP 数据通信。本节重点介绍 TCP 协议的网络编程方法。

10.1.1 TCP/IP 协议基础

TCP（Transmission Control Protocol，传输控制协议）是基于连接的通信协议。在互联网设备进行通信之前，数据交流的双方必须建立安全可靠的连接。这一过程类似于打电话一样，在拨通电话后，先说一声"喂"，等对方确认了再开始正常通话。TCP 协议使用了重传机制，在接收端接收到发送端发送的一个报文之后，会返回响应报文，如果发送端没有收到这个响应报文，则会继续重发，这样即便在网络拥塞时，也不会出现传输错误的情况。

在 Java 中，JDK 提供了 TCP、UDP 网络通信的 API，其中的 Socket，ServerSocket 两个类用来建立TCP/IP 协议的网络通信，Socket 本质上就是对传输层中的 TCP 协议进行了的封装，由于 TCP 是端到端的通信，要实现 Socket 的传输，就要构建客户端与服务端。TCP 协议通信原理图如图 10.1 所示。

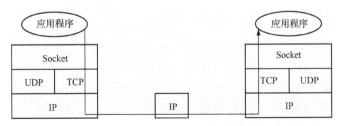

图 10.1 TCP 协议通信原理图

10.1.2 Socket 与 ServerSocket 建立通信

在介绍 Socket 与 ServerSocket 通信之前，需要了解数据通信中涉及的一些异常类型，这些异常类型在后文中将会用到。异常类型如表 10.1 所示。

表 10.1 异常类型

异常类型	描述
UnkownHostException	主机名字或 IP 错误
ConnectException	服务器拒绝连接、服务器没有启动（超出队列数，拒绝连接）
SocketTimeoutException	连接超时
BindException	Socket 对象无法与指定的本地 IP 地址或端口绑定

在如图 10.1 所示的 TCP 协议通信原理图中，由于两个通信实体已经连接，所以并没有区分客户端与服务端，但在两个通信实体通信之前，其中作为服务端的一方需要建立等待连接的机制，等待客户端的请求，这时就需要 ServerSocket 对象监听来自客户端的 Socket 连接，如果没有连接，将会一直处于等待状态。Socket 与 ServerSocket 的交互原理如图 10.2 所示。

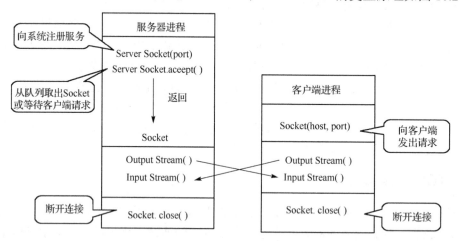

图 10.2 Socket 与 ServerSocket 的交互原理

如图 10.2 所示，完成 Socket 通信首先需要建立服务端，ServerSocket 类提供了如表 10.2 所示的构造方法，用于创建 ServerSocket 对象。

表 10.2 ServerSocket 类的构造方法

方法	描述
ServerSocket(int port) throws IOException	创建一个指定端口的 ServerSocket，端口范围为 0～65535
ServerSocket(int port, int backlog) throws IOException	增加 backlog 参数用于改变连接队列长度
ServerSocket(int port, int backlog, InetAddress bindAddr) throws IOException	增加 localAddr 参数用于将 ServerSocket 绑定到指定的 IP 地址

在指定了服务端之后，便需要监听来自客户端的连接请求，ServerSocket 提供了 accept() 方法。如果服务端接收到了客户端的 Socket 连接请求，该方法将返回一个与连接客户端 Socket 对应的 Socket，正如图 10.1 所示，TCP 通信双方都需要有一个 Socket。

在使用 ServerSocket 时需要注意以下几点：

- 如果端口被占用或者没有权限使用某些端口，则会抛出 BindException 错误。例如，1-1023 端口需要管理员权限。
- 如果设置端口为 0，则系统会自动为其分配一个端口。
- ServerSocket 一旦绑定了监听端口，就无法更改。ServerSocket()可以实现在绑定端口前设置其他的参数。

通常，在使用完 ServerSocket 之后，调用 ServerSocket 的 close()方法关闭该 ServerSocket。下面的代码展示了如何运用 ServerSocket 建立服务端：

```
ServerSocket serverSocket = new ServerSocket(8888);//设置服务端的端口为8888
while(true){
try{
    Socket socket=serverSocket.accept();
                //从连接队列中取出一个连接，如果没有则等待
System.out.println("新增连接: "+socket.getInetAddress()+":"+socket.getPort());
        .../ /接收和发送数据
}catch(IOException e){e.printStackTrace();}
finally{
    try{
            if(socket!=null)
            socket.close();               //与一个客户端通信结束后，要关闭Socket
        }catch(IOException e){e.printStackTrace();}
    }
}
```

注意：由于服务端通常运行在有固定 IP 地址的服务器上，所以以上代码可以直接运行在 PC 上，将 PC 作为服务端。

建立好服务端之后，就可以开始建立客户端，通常使用 Socket 构造函数直接指定需要连接的服务器 IP 地址，Socket 提供了如表 10.3 所示的几个构造方法。

表 10.3 Socket 类的构造方法

方法	描述
Socket(InetAddress address/String host, int port) throws UnknownHostException, IOException	创建连接到指定的远程 IP 地址及端口的 Socket，未指定本地 IP 地址及端口，则采用本地主机的 IP 地址
Socket(InetAddress address/String host, int port, InetAddress localAddress/String host, int localPort) throws IOException	对比于上面的方法，指定了本地的 IP 地址及端口

Socket 类还提供了一些常用于建立通信连接的方法，如表 10.4 所示。

表 10.4 Socket 类的常用方法

返回类型	方法	描述
InetAddress	getInetAddress()	获取远程服务端的 IP 地址
int	getPort()	获取远程服务端的端口
String	getLocalAddress()	本地客户端的 IP 地址
int	getLocalPort()	本地客户端的端口
InputString	getInputStream()	获得输入流
OutputString	getOutputStream()	获得输出流
void	connect(SocketAddress remoteAddr)	连接到指定的 IP 地址
void	connect(SocketAddress remoteAddr, int timeout)	连接到指定的 IP 地址，如果时间超出 timeout 则抛出超时异常

在上述方法中，getInputString()方法与getOutString()方法最为重要，在通信时获取输入流与输出流，下面的例子展示了如何运用Socket类建立，实现连接到指定IP端口并获取服务端发送的数据的功能。该程序创建连接后，将接收到的数据显示在文本框中。

【例10.1.1】 建立客户端程序实例

```java
package com.example.socketclient;
import android.os.Bundle;
import android.support.v7.app.AppCompatActivity;
import android.widget.TextView;
import java.io.BufferedReader;
import java.io.IOException;
import java.io.InputStreamReader;
import java.net.Socket;
import java.net.SocketTimeoutException;
public class ClientActivity extends AppCompatActivity {
    TextView tvInfo;
    @Override
    protected void onCreate(Bundle savedInstanceState) {
        super.onCreate(savedInstanceState);
        setContentView(R.layout.activity_client);
        tvInfo = (TextView)findViewById(R.id.receivedInfo);
        new Thread() {
            public void run() {
                try{
                    Socket socket =
                        new Socket("192.168.252.1", 4666); //创建Socket
                    socket.setSoTimeout(3000);      //超时设置为3s
                    BufferedReader br = new BufferedReader(new
                        InputStreamReader(socket.getInputStream()));
                    String lien = br.readLine();   //读取接收到的数据
                    tvInfo.setText("Client received: " + lien);
                    br.close();
                    socket.close();
                }
                catch (SocketTimeoutException e){
                    e.printStackTrace();           //抛出连接超时异常
                }
                catch(IOException e){
                    e.printStackTrace();
                }
            }
        }.start();
    }
}
```

注意：上面的代码为socket通信开了一个新的线程，这是由于进行网络通信的这类不稳定耗时的任务，可能对UI线程造成阻塞，所以Android 2.3之后的平台都不允许在UI线程中建立网络连接以及数据通信。

可以看到笔者本地的 IP 配置为"192.168.252.1",读者可以根据自己的局域网设置进行配置,由于使用了网络连接,还需要注意在 Mainifest 文件中添加网络访问权限,如下所示:

```
<uses-permission android:name="android.permission.INTERNET"/>
```

接着编写服务端的代码,可以直接用 Java 编写服务端的代码,然后在 PC 上运行。在发送数据流时,为了防止乱码,应注意采用"utf-8"字符集,代码如下所示。

【例 10.1.2】 建立服务端程序实例

```java
package com.example.socketserver;
import java.io.IOException;
import java.io.OutputStream;
import java.net.ServerSocket;
import java.net.Socket;
public class Server {
    public static void main(String[] args) throws IOException{
        //TODO Auto-generated method stub
        ServerSocket serverSocket = new ServerSocket(4666);  //指定端口
        while(true){
            Socket socket = serverSocket.accept();  //获取socket连接请求
            OutputStream os = socket.getOutputStream();  //获取输出流
            os.write("This is Server! ".getBytes("utf-8"));  //发送数据流
            os.close();
            socket.close();
        }
    }
}
```

以上的代码在 PC 端运行后,在 Android 端运行客户端程序,即可看到程序运行结果如图 10.3 所示。

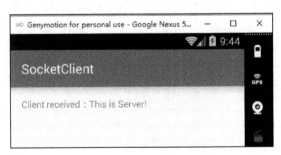

图 10.3 程序运行结果

10.2 使用 URL 访问网络

10.2.1 使用 URL 获取网络资源

URL(Uniform Resource Locator)对象代表统一资源定位器,它是指向互联网"资源"的指针。资源可以是简单的文件或目录,也可以是对更复杂的对象的引用,例如对数据库或搜索引擎的查询。通常，URL 可以由协议名、主机、端口和资源组成，即满足如下格式：protocol://host:port/resourceName。例如以下的 URL 地址：

http://www.cqupt.com/index.php

URL 类提供了多个用于创建 URL 对象的构造方法，一旦获得了 URL 对象之后，可以调用如表 10.5 所示的常用方法来访问该 URL 对应的资源。

表 10.5 访问 URL 对应资源的常用方法

类型	方法	简述
String	getFile()	获取此 URL 的资源名
String	getHost()	获取此 URL 的主机名
String	getPath()	获取此 URL 的路径部分
String	getProtocol()	获取此 URL 的协议名称
String	getQuery()	获取此 URL 的查询字符串部分
String	getPort()	获取此 URL 的端口号
URLConnection	openConnection()	返回一个 URLConnection 对象，它表示到 URL 所引用的远程对象的连接
InputStream	openStream()	打开与此 URL 的连接，并返回一个用于读取该 URL 资源的 InputStream

在上一节中已经介绍过，在 Android 2.3 之后的平台中，网络访问不能在 UI 线程中进行，所以本节将介绍如何运用 AsyncTask 类创建异步线程并访问 URL 网络资源。

首先了解 AsyncTask 类的基本知识。AsyncTask 类的特点是任务在主线程之外运行，而回调方法是在主线程中执行，这就有效地避免了使用 Handler 带来的麻烦。AsyncTask 类是抽象类，其中定义了如下三种泛型类型。

● Params 启动任务执行的输入参数，比如网络请求的 URL。
● Progress 后台任务执行的百分比，如果不需要可以定义为 Void 类型。
● Result 后台执行任务最终返回的结果，比如 String。

AsyncTask 抽象出后台线程运行的五个状态，分别是准备运行、正在后台运行、进度更新、完成后台任务、取消任务。对应于这几个状态，AsyncTask 类提供了如表 10.6 所示的常用方法。

表 10.6 AsyncTask 提供的常用方法

序号	方法	简述
1	onPreExecute()	该方法将在执行实际的后台操作前被 UI thread 调用。可以在该方法中做一些准备工作，如在界面上显示一个进度条
2	doInBackground(Params...)	将在 onPreExecute 方法执行后马上执行，该方法运行在后台线程中。主要负责执行耗时的后台计算工作。可以调用 publishProgress 方法来更新实时的任务进度。该方法是抽象方法，子类必须实现
3	onProgressUpdate(Progress...)	在 publishProgress 方法被调用后，UI thread 将调用这个方法从而在界面上展示任务的进展情况，例如通过一个进度条进行展示
4	onPostExecute(Result)	在 doInBackground 执行完成后，onPostExecute 方法将被 UI thread 调用，后台的计算结果将通过该方法传递到 UI thread
5	onCancelled()	取消任务

注意：doInBackground 方法和 onPostExecute 的参数必须对应，这两个参数在 AsyncTask 声明的泛型参数列表中指定，第一个为 doInBackground 接收的初始化参数，第二个为显示进度的参数，第三个为 doInBackground 返回和 onPostExecute 得到的结果参数。

下面用一个例子来展示异步线程在实际场景中的应用。在这一个例子中将会加载网络中的一张图片，读者可以通过百度图片搜索，右键单击图片获取图片地址就可以获取到该图片的 URL 资源。

【例 10.2】 访问 URL 网络资源实例

```java
package com.example.url;
import android.graphics.Bitmap;
import android.graphics.BitmapFactory;
import android.os.AsyncTask;
import android.support.v7.app.AppCompatActivity;
import android.os.Bundle;
import android.widget.ImageView;
import java.io.InputStream;
import java.net.MalformedURLException;
import java.net.URL;
public class WebActivity extends AppCompatActivity {
    ImageView imgView;
    Bitmap bm;
    @Override
    protected void onCreate(Bundle savedInstanceState) {
        super.onCreate(savedInstanceState);
        setContentView(R.layout.activity_web);
        imgView = (ImageView)findViewById(R.id.imgView);
        ShowPicture("http://pic.to8to.com/attch/day_160218/201602
                18_d968438a2434b62ba59dH7q5KEzTS6OH.png");
    }
    public void ShowPicture(String url){
        new AsyncTask<String, Void , String>() {
            @Override
            protected String doInBackground(String... params) {
                try{
                    URL url = new URL(params[0]);          //定义url为第一个参数
                    InputStream is = url.openStream();     //获取输入流
                    bm = BitmapFactory.decodeStream(is);   //解析图片
                    is.close();                            //关闭输入流
                }catch (MalformedURLException e){
                    e.printStackTrace();
                }catch (Exception e){
                    e.printStackTrace();
                }
                return null;
            }
            @Override
            protected void onPostExecute(String s) {
                super.onPostExecute(s);
                imgView.setImageBitmap(bm);                //显示图片
            }
        }.execute(url);
    }
}
```

同样,上述代码需要在 Mainifest 文件中添加网络访问权限,如下所示:

```
<uses-permission android:name="android.permission.INTERNET"/>
```

程序运行结果如图 10.4 所示。

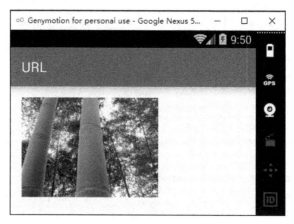

图 10.4　程序运行结果

10.2.2　使用 URLConnection 提交请求

在了解了 URL 实现一些简单的操作之后,可以使用 URL 中所引用远程对象的连接 URLConnection,URLConnection 对象可以向所代表的 URL 发送请求和读取 URL 的资源。通常,创建一个和 URL 的连接需要如下几个步骤。

- 创建 URL 对象,并通过调用 openConnection 方法获得 URLConnection 对象。
- 设置 URLConnection 参数和普通请求属性。
- 向远程资源发送请求。
- 远程资源变为可用,程序可以访问远程资源的头字段和通过输入流来读取远程资源返回的信息。
- 关闭输入流。

注意:在连接到远程资源之前必须对 URLConnection 进行配置,并且 URLConnection 实例不能重用,每一个连接都需要运用不同的 URLConnection 实例。

URLConnection 类提供了如表 10.7 所示的常用方法,用于在建立连接之前设置请求头字段。

表 10.7　URLConnection 类提供的常用方法

类型	方法	简述
InputStream	getInputStream()	获取该 URLConnection 的输入流
OutputStream	getOutputStream()	获取该 URLConnection 的输出流
void	setDoInput(boolean newValue)	设置该 URLConnection 是否允许输入
void	setDoOutput(boolean newValue)	设置该 URLConnection 是否允许输出
void	setReadTimeout(int timeoutMillis)	设置连接期间的最大等待时间
void	setRequestProperty	设置指定的请求头字段的值
void	setUseCaches(boolean newValue)	设置标志以说明该连接是否允许使用缓存

通常，网络请求分为两类：get 请求方式和 post 请求方式。这两种请求方式的区别如下。
- get 请求方式，只需要使用 connect()方法建立和远程的通信，参数传递时只需显式地将参数添加在 URL 之后，也就是说在构造 URL 时就需要添加参数，代码如下所示：

```
String urlName = url + "?" + param;              //定义URL
URL realUrl = new URL(urlName);
URLConnection conn = realUrl.openConnection();   //打开和URL之间的连接
conn.setRequestProperty("accept", "*/*");        //设置通用的请求属性
……                                               //设置通用的请求属性
conn.connect();
```

- post 请求方式，URL 中不需添加参数则，要在使用 connect()方法之后，获取 URLConnection 对象所对应的输出流来发送网络请求的参数，代码如下所示：

```
URL realUrl = new URL(url);                      //打开和URL之间的连接
URLConnection conn = realUrl.openConnection(); conn.setRequestProperty
        ("accept", "*/*");
……                                               //设置通用的请求属性
conn.setDoOutput(true);
conn.setDoInput(true);
out = new PrintWriter(conn.getOutputStream());
                                                 //获取URLConnection对象对应的输出流
out.print(param);                                //发送请求参数
```

post 请求方式较常用，原因如下。
- get 是从服务器上获取数据，post 是向服务器传送数据。
- get 传送的数据量较小，不能大于 2KB。post 传送的数据量较大，一般被默认为不受限制。
- 虽然 get 执行效率比 post 高，但是 get 安全性比 post 高。

下面用一个例子展示如何运用 URLConnection 提交网络请求。由于该例程需要后台服务器的支持，所以笔者编写了一个 JavaWeb 应用，然后将笔记本作为服务器，这样才能实现网络请求。笔者这里只是简单地介绍了 JavaWeb 工程的创建，对于如何搭建测试环境，读者可以查阅相关的教程。

1）创建服务器端

读者可以使用 MyEclipse 创建 Web 工程，用于模拟服务器端的 Web 应用，Web 工程名为"MyServer"，修改 Index.jsp 文件，创建两个文本输入框以及一个提交按钮，作为登录显示页。

【例 10.3】 JavaWeb 服务器端实例

使用 URLConnection 提交请求，服务端代码如下：

```
<%@ page language="java" import="java.util.*" pageEncoding="UTF-8"%>
<html>
  <head>  <title>My JSP 'index.jsp' starting page</title>  </head>
  <body> 登录演示 <br>
      <form action="${pageContext.request.contextPath}
                    /servlet/LoginServlet" method="post">
        账号：<input type = "text" name = "username" ><br>
```

```
            密码: <input type = "text" name = "password" ><br>
            <input type = "submit" value = "提交">
        </form>
    </body>
</html>
```

运行 Tomcat，在浏览器中访问"localhost/Myserver"便可以看到如图 10.5 所示的页面，表示可以成功运行。

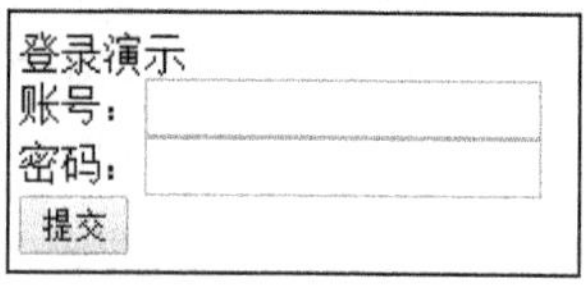

接着在该 Web 工程中，创建一个继承自 HttpServlet 的 LoginServlet 类，并实现其中的 doPost()方法，用来响应如图 10.5 所示页面的用户操作。具体实现如下：

图 10.5 页面显示效果

```java
package test;
import java.io.IOException;
import java.io.PrintWriter;
import javax.servlet.ServletException;
import javax.servlet.http.HttpServlet;
import javax.servlet.http.HttpServletRequest;
import javax.servlet.http.HttpServletResponse;
public class LoginServlet extends HttpServlet{
public void doPost(HttpServletRequest request, HttpServletResponse response)
            throws ServletException, IOException {
    response.setContentType("text/html;charset=utf-8");
    request.setCharacterEncoding("utf-8");
    response.setCharacterEncoding("utf-8");
    PrintWriter out = response.getWriter();
    String username = request.getParameter("username");  //获取账号
    String password = request.getParameter("password");  //获取密码
    //判断账号密码是否正确
    if(username.equals("admin") && password.equals("123456")) {
        out.print("Login succeeded!");
    }else {
        out.print("Login failed!");
    }
    out.flush();
    out.close();
  }
}
```

在 Tomcat 中运行上述代码，若在如图 10.5 所示的页面中输入账号为"admin"、密码为"123456"，单击"提交"按钮，则会得到"Login succeeded!"的提示信息，如图 10.6 所示，这样我们 Web 端的环境就配置成功了。

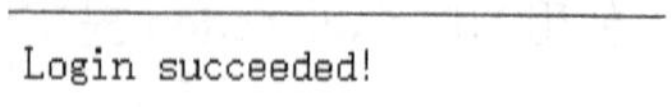

Login succeeded!

2) 客户端

图 10.6 网页登录效果

在客户端首先需要创建一个登录界面，该界面包含两个文本输入框、一个"提交"按钮、一个返回结果的文本框。然后，在程序中使用异步任务提交网络请求，这里采用 post 请求方式，代码如下所示。

【例 10.4.1】 Android 客户端 URLConnection 请求实例

使用 URLConnection 提交请求，代码如下：

```java
package com.example.urlconnection;
import android.os.AsyncTask;
import android.support.v7.app.AppCompatActivity;
import android.os.Bundle;
import android.view.View;
import android.widget.Button;
import android.widget.EditText;
import android.widget.TextView;
import java.io.BufferedReader;
import java.io.IOException;
import java.io.InputStreamReader;
import java.io.PrintWriter;
import java.net.MalformedURLException;
import java.net.URL;
import java.net.URLConnection;
public class URLConnectionDemo extends AppCompatActivity {
    EditText et_username,et_password;
    Button btn_submit;
    TextView tv_result;
    @Override
    protected void onCreate(Bundle savedInstanceState) {
        super.onCreate(savedInstanceState);
        setContentView(R.layout.activity_urlconnection_demo);
        et_username = (EditText)findViewById(R.id.ed_username);
        et_password = (EditText)findViewById(R.id.ed_password);
        btn_submit = (Button)findViewById(R.id.btn_submit);
        tv_result=(TextView)findViewById(R.id.tv_result);
        btn_submit.setOnClickListener(new View.OnClickListener() {
            @Override
            public void onClick(View v) {
                loginCheck(et_username.getText().toString()
                        ,et_password.getText().toString());
            }
        });
    }
    public void loginCheck(String username,String password)
    {
        final String url =
            "http://172.0.0.1:8888/MyServer/Servlet/LoginServlet";//访问的url
        final String param = "username=" + username + "&"
                        + "password=" + password;        //设置参数
        new AsyncTask<String, Void, String>(){
            @Override
            protected String doInBackground(String... params) {
                PrintWriter out = null;
                BufferedReader in = null;
```

```java
            String result = null;
            try {
                URL url = new URL(params[0]);
                URLConnection conn = url.openConnection();
                conn.setRequestProperty("accept","/*");
                                        //设置URLConnection参数
                conn.setRequestProperty("connection","Keep-Alive");
                conn.setRequestProperty("user-agent",
            "Mozilla/4.0 (compatible; MSIE 6.0; Windows NT 5.1; SV1)");
                conn.setDoInput(true);      //允许输入
                conn.setDoOutput(true);     //允许输出
                out = new PrintWriter(conn.getOutputStream());
                out.print(param);           //设置参数
                in = new BufferedReader(new InputStreamReader(conn
                                .getInputStream()));
                String line;
                while((line= in.readLine())!=null)
                {
                    result += line;         //获取返回结果
                }
            } catch (MalformedURLException e) {
                e.printStackTrace();
            } catch (IOException e) {
                e.printStackTrace();
            }
            return result;
        }
        @Override
        protected void onPostExecute(String result) {
            if(result!=null){
                tv_result.setText("验证结果: " + result); //显示返回结果
            }
            super.onPostExecute(result);
        }
    }.execute(url);
}
```

同样，上述代码需要在Mainifest文件中添加网络访问权限，如下所示：

```
<uses-permission android:name="android.permission.INTERNET"/>
```

注意：在提交参数时，参数的格式为"param1=value&param2=value…"，中间的"&"用于分隔参数，不能省去。

程序运行结果如图10.7所示。

第 10 章 Android 的网络编程基础   201

图 10.7 程序运行结果

10.2.3 使用 HttpURLConnection 实现网络通信

在上一节中，展示了如何使用 URLConnection 提交请求，这一节将介绍使用 HttpURLConnection 访问网络。HttpURLConnection 继承自 URLConnection，新增了一些用于操作 Http 资源的方法，如表 10.8 所示。

表 10.8 HttpURLConnection 类提供的常用方法

类型	方法	简述
Int	getResponseCode	获取服务器的响应代码
String	getResponseMessage	获取服务器的响应消息
String	getRequestMethod	获取发送请求的方法
void	setRequestMethod（String method）	设置发送请求的方法

下面将采用 HttpURLConnetion 方法实现之前提到的网络请求，只需要对异步请求的代码做如下修改。

【例 10.4.2】 Android 客户端 HttpURLConnetion 请求实例

使用 HttpURLConnection 提交请求，代码如下：

```
public void loginCheck2(String username,String password)
{
    final String url = "http://172.0.0.1:8888/MyServer/Servlet/
                       LoginServlet";                //访问的url
    final String param = "username=" + username + "&"
                       + "password=" + password;    //设置参数
    new AsyncTask<String, Void, String>(){
        @Override
        protected String doInBackground(String... params) {
            PrintWriter out = null;
            BufferedReader in = null;
            String result = null;
```

```java
        try {
            URL url = new URL(params[0]);
            HttpURLConnection httpConn =
                    (HttpURLConnection) url.openConnection();
                    conn.setRequestProperty("accept","/*");
                                //设置URLConnection参数
            conn.setRequestProperty("connection","Keep-Alive");
            conn.setRequestProperty("user-agent",
   "Mozilla/4.0 (compatible; MSIE 6.0; Windows NT 5.1; SV1)");
            httpConn.setDoInput(true);          //允许输入
            httpConn.setDoOutput(true);         //允许输出
            httpConn.setRequestMethod("post"); //设置为POST请求方式
            httpConn.setUseCaches(false);   //使用Post方式不能使用缓存
            out = new PrintWriter(httpConn.getOutputStream());
            out.print(param);                   //设置参数
            int responseCode = httpConn.getResponseCode();
            if(responseCode==HttpURLConnection.HTTP_OK){
                in = new BufferedReader(new
                    InputStreamReader(httpConn.getInputStream()));
                String line;
                while((line= in.readLine())!=null)
                {
                    result += line;             //获取返回结果
                }
            }
        }
        catch (MalformedURLException e) {
            e.printStackTrace();
        } catch (IOException e) {
            e.printStackTrace();
        }
        return result;
    }
    @Override
    protected void onPostExecute(String result) {
        if(result!=null){
            tv_result.setText("验证结果: " + result); //显示返回结果
        }
        super.onPostExecute(result);
    }
}.execute(url);
}
```

运行上述代码同样可以得到如图10.7所示的结果。可能有的读者在以前学习Android网络时接触过HttpClient这个类,但是Goolge在Android API23之后就不在SDK中自带HttpClient了,原因是Apache不再维护HttpClient了,官方推荐使用HttpURLConnection。如果仍需要使用，最便捷的方法是在 build.gradle 中的 android {}中加上 useLibrary"org.apache.http.legacy" 即可，有需要的读者可以自行尝试。

10.3 使用 WebView

随着后台技术的不断发展，App 前端的应用都内置了 Web 页面的界面，这个界面就是由 WebView 组件渲染出来的。这样在开发一些类似于帖子、活动页面的功能时，都使用后台统一开发，而不同的前端只需直接调用即可，极大地优化了开发效率。WebView 有如下几个优点：可以直接显示和渲染 Web 页面或者网页；可以直接调用网络上或者本地的 html 文件，也可以和 JavaScript 交互调用。

10.3.1 使用 WebView 浏览网页

WebView 组件类似于一个小型的浏览器，提供了一些类似于浏览器的常用方法，如表 10.9 所示。

表 10.9 WebView 类提供的常用方法

类型	方法	描述
void	goBack()	返回
void	goForward()	前进
void	getProgress()	得到访问进度
void	getTitle()	得到当前访问页的标题
void	loadUrl(String url)	加载指定的页面
void	reload()	重新加载当前页面
void	savePassword(String host, String username, String password)	保存密码
boolean	zoomIn()	放大网页
boolean	zoomOut()	缩小网页
WebSetting	getSettings()	返回 WebSetting 对象

下面利用 WebView 做一个迷你浏览器，在布局界面中放入"后退"、"前进"、"刷新"、"进入"按钮，以及网址输入框，对布局界面不再赘述，主程序代码如下：

【例 10.5】 使用 WebView 加载网页视图实例

```
package com.example.webview;
import android.support.v7.app.AppCompatActivity;
import android.os.Bundle;
import android.view.KeyEvent;
import android.view.View;
import android.webkit.WebSettings;
import android.webkit.WebView;
import android.webkit.WebViewClient;
import android.widget.Button;
import android.widget.EditText;
public class MiniBrowser extends AppCompatActivity implements
                        View.OnClickListener{
    Button btnBack,btnForward,btnRefresh,btnGo;
```

```java
    EditText etURL;
    private WebView webView;
    @Override
    protected void onCreate(Bundle savedInstanceState) {
        super.onCreate(savedInstanceState);
        setContentView(R.layout.activity_mini_browser);
        btnBack = (Button)findViewById(R.id.btnBac);
        btnForward = (Button)findViewById(R.id.btnFow);
        btnRefresh = (Button)findViewById(R.id.btnRef);
        btnGo = (Button)findViewById(R.id.btnGo);
        etURL = (EditText)findViewById(R.id.etUrl);
        webView = (WebView)findViewById(R.id.webView);
        btnBack.setOnClickListener(this);
        btnForward.setOnClickListener(this);
        btnRefresh.setOnClickListener(this);
        btnGo.setOnClickListener(this);
    }
    @Override
    public void onClick(View v) {
        switch (v.getId()){
            case R.id.btnBac:
                webView.goBack();
                break;
            case R.id.btnFow:
                webView.goForward();
                break;
            case R.id.btnRef:
                webView.reload();
                break;
            case R.id.btnGo:
                if(etURL.getText().toString()!=null){
                    //加载需要显示的网页
                    webView.loadUrl("http://"+etURL.getText().toString());
                    //设置Web视图
                    webView.setWebViewClient(new webViewClient());
                }
                break;
            default: break;
        }
    }
    @Override
    //设置回退
    public boolean onKeyDown(int keyCode, KeyEvent event) {
        if ((keyCode == KeyEvent.KEYCODE_BACK) && webView.canGoBack()) {
            webView.goBack(); //goBack()表示返回WebView的上一页面
            return true;
        }
```

```
            return false;
        }
        //Web 视图
        private class webViewClient extends WebViewClient {
            public boolean shouldOverrideUrlLoading(WebView view, String url) {
                view.loadUrl(url);
                return true;
            }
        }
    }
}
```

程序运行结果如图 10.8 所示，类似于一个迷你浏览器，输入网址后，单击"前往"按钮即可看到下方的 WebView 组件加载出网页，单击上方的"后退"、"前进"与"刷新"可以执行对应的操作。

图 10.8　程序运行结果

10.3.2　加载本地 html 页面

在应用的实际开发过程中，某些界面可以直接加载 html 页面，例如注册说明或者使用帮助界面，这样可以一次开发后直接使用多个平台，十分便捷。下面介绍如何在本地工程中添加 html 文件，然后在程序中调用显示。

首先需要在工程目录下创建 assets 文件目录，用于存放 html 资源，创建方法为：右键单击工程，在弹出的菜单中依次单击"New"→"Folder"→"Assets Folder"，如图 10.9 所示。

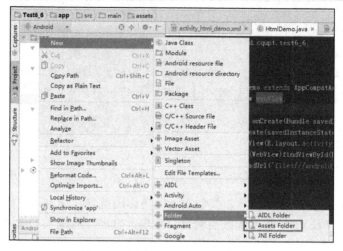

图 10.9　创建 Assets 文件目录

然后使用文本编辑器编写 html 文件，代码如下所示：

```
<!DOCTYPE html>
<html>
    <head>
        <meta http-equiv="Content-Type" content="text/html; charset=GBK">
        <title>Insert title here</title>
    </head>
    <body >
        加载本地 HTML 网页！
    </body>
</html>
```

注意：为防止乱码，字符集选择 GBK 格式。然后在布局界面中插入 WebView 组件，最后定义主程序，如下所示。

【例 10.6】　使用 WebView 加载本地 html 实例

```
package com.example.webviewlocal;
import android.support.v7.app.AppCompatActivity;
import android.os.Bundle;
import android.webkit.WebView;
public class HtmlDemo extends AppCompatActivity {
    private WebView webView;
    @Override
    protected void onCreate(Bundle savedInstanceState) {
        super.onCreate(savedInstanceState);
        setContentView(R.layout.activity_html_demo);
        webView = (WebView)findViewById(R.id.webview);
        webView.loadUrl("file:///android_asset/index.html");
                    //加载本地 html 页面
    }
}
```

程序运行结果如图 10.10 所示。

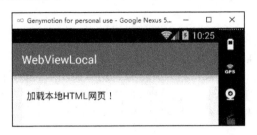

图 10.10　程序运行结果

10.3.3　JavaScript 交互调用

上述加载本地 html 文件的示例只是加载了界面视图，但涉及更多复杂的页面就无法加载了。html 文件中通常带有 JavaScript 脚本。细心的读者可能发现在加载网页视图时，由于没有对 WebView 进行设置，在某些界面会看到如图 10.11 所示的界面。

图 10.11　加载 html 页面

这时，就需要用 WebSetting 类对 WebView 进行设置以开启对 JavaScript 的支持，通过 WebView 提供的 getSettings()方法可以获取该对象。在表 10.10 中展示了 WebSetting 类提供的常用方法。

表 10.10　WebSetting 类提供的常用方法

类型	方法	简述
void	setAllowFileAccess((boolean allow))	设置可以访问文件
void	setBuiltInZoomControls(boolean enabled)	设置是否支持缩放操作
void	setBlockNetworkImage(boolean flag)	设置是否显示网络图像
void	setJavaScriptEnabled(boolean enabled)	设置是否启用 JavaScript 支持
void	setDefaultFontSize(int size)	设置默认字体大小
void	setDefaultTextEncodingName(String encoding)	设置默认字符集格式
void	setGeolocationEnabled(boolean enabled)	设置是否可以获取地理位置信息
void	getTitle()	得到当前访问页的标题

所以为了开启对 JavaScript 的支持，在例 10.5 中还需要在 loadUrl() 方法之前设置 WebView 属性，方法如下所示：

```
WebSettings webSettings = webview.getSettings();
//设置 WebView 属性，能够执行 JavaScript 脚本
webSettings.setJavaScriptEnabled(true);
//设置可以访问文件
webSettings.setAllowFileAccess(true);
//设置支持缩放
webSettings.setBuiltInZoomControls(true);
```

在 http 页面中植入 JavaScript 最大的好处是能和 Android 进行交互操作，如在 App 中弹出 Toast，或者展示列表，或者调用手机的通话功能，直接拨打页面上的电话号码。下面用一个例子展示 JavaScript 调用 Android 的通话功能。

【例 10.7】 JavaScript 与 Android 交互实例

定义 html 页面如下：

```
<!DOCTYPE html>
<html>
<head>
    <meta http-equiv="Content-Type" content="text/html; charset=UTF-8">
    <title>Insert title here</title>
</head>
    <body >
    单击呼叫电话
    <a href="javascript:demo.call('10086')">10086</a>  <!--这是一个超链接，
            调用了 Android 代码中的 call()方法-->
    </body>
</html>
```

Android 端代码如下：

```
package com.example.javascript;
import android.Manifest;
import android.content.Intent;
import android.content.pm.PackageManager;
import android.net.Uri;
import android.support.v4.app.ActivityCompat;
import android.support.v7.app.AppCompatActivity;
import android.os.Bundle;
import android.webkit.JavascriptInterface;
import android.webkit.WebSettings;
import android.webkit.WebView;
public class WebViewCall extends AppCompatActivity {
    private WebView webView;
    @Override
    protected void onCreate(Bundle savedInstanceState) {
        super.onCreate(savedInstanceState);
        setContentView(R.layout.activity_web_view_call);
```

```
            webView = (WebView) findViewById(R.id.webview);
            WebSettings ws = webView.getSettings();
            ws.setJavaScriptEnabled(true);                     //启动 JavaScript
            ws.setDefaultTextEncodingName("UTF-8");            //定义字符集为"UTF-8"
            webView.loadUrl("file:///android_asset/call.html");//加载 html 页面
            webView.addJavascriptInterface(this,"demo");//html 页面中的 JavaScript
        }
    @JavascriptInterface
                //一定要添加这个标签，才能将该方法暴露给 javaScript
        public void call(final String phone) {
            if (ActivityCompat.checkSelfPermission(this    //检测权限
                    , Manifest.permission.CALL_PHONE) !=
                    PackageManager.PERMISSION_GRANTED) {
                return;}
            Intent intent = new Intent(Intent.ACTION_CALL, Uri.parse
                        ("tel:" + phone));
            startActivity(intent);
        }
    }
```

注意：在上面代码中，需要暴露给 JavaScript 的方法必须在之前加上"@JavascriptInterface"，而且要在"Manifest"文件中添加通话权限，如下所示：

```
<uses-permission android:name="android.permission.CALL_PHONE"/>
```

程序运行结果如图 10.12 所示。

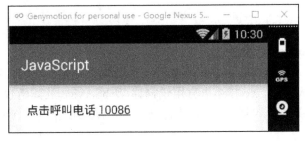

图 10.12　程序运行结果

10.4　使用 WebService 进行网络编程

10.4.1　WebService 基础

前面介绍了多种与后台服务端通信的操作，但是在那种开发模式下会存在平台的约束，对于一个支持多平台的后台业务中心来说，WebService 是不错的解决方案，例如上述 JavaWeb 开发的网络程序只适用于 Java 编写的客户端显示。如果一个应用程序需要跨平台操作，则可以采用 WebService 作为后台服务端，采用 SOAP（简单对象访问协议）进行通信，通过统一的 XML 语言传输数据。

WebService 简单来说就是一个网络应用程序，它向外部暴露了能够通过 Web 调用的 API 接口。因为需要实现互操作性，WebService 平台需要一套可创建分布式应用程序的协议，该平台具有以下三大核心技术。

- **XML+XSD**

WebService 采用了 HTTP 协议传输数据，采用 XML 封装数据，XML 的优点在于易于创建、便于分析以及有着很好的平台无关性，任何平台都可以轻松地解析数据，而 XML 没有定义一套标准的数据类型，正好与 XSD 相结合，解决了数据类型的定义问题。

- **SOAP**（简单对象访问协议）

SOAP 是交换数据的一种协议规范，是一种轻量级的基于 XML 的协议，其数据传输可以通过 HTTP 或 SMTP 协议，SOAP 消息通常都是从发送端到接收端的单向传输，一条完整的 SOAP 消息包含：

(1)必需的 Envelope 元素，可把此 XML 文档标识为一条 SOAP 消息。
(2)可选的 Header 元素，包含头部信息。
(3)必需的 Body 元素，包含所有的调用和响应信息。
(4)可选的 Fault 元素，提供有关在处理此消息所发生错误的信息。

其格式如下所示：

```
<SOAP-ENV:Envelope
    各种属性>
<!--示例-->
    <SOAP:HEADER>
    </SOAP:HEADER>
    <SOAP:Body>
    </SOAP:Body>
</SOAP-ENV:Envelope>
```

- **WSDL**（WebService 描述语言）

当一个 WebService 完成后，若想让其他开发人员调用其中的功能，则需要一定的规范性操作。基于 XML 的 WebService 描述语言用于描述 WebService 及其函数、参数与返回值，有些软件能够直接获取或生成 WSDL。这是十分方便的，WSDL 描述了 WebService 的三个基本属性：

(1)服务所提供的操作。
(2)如何访问服务。
(3)服务位于何处(通过 URL 来确定)。

10.4.2 调用 WebService

Android 平台调用 WebService 需要依赖于第三方类库 ksoap2-android。ksoap2-android 是 Google 为 Android 平台开发 WebService 客户端提供的开源项目，所以在开发之前，需要将 ksoap2-android 的 jar 包添加到 Android 工程中，步骤如下。

(1)在 Google 提供的项目下载网站中下载开发包，网址为"https://code.google.com/archive/p/ksoap2-android/source"，单击"Downloads"，选择"ksoap2-android-assembly- 2.4-jar-with-dependencies.jar"，如图 10-13 所示。

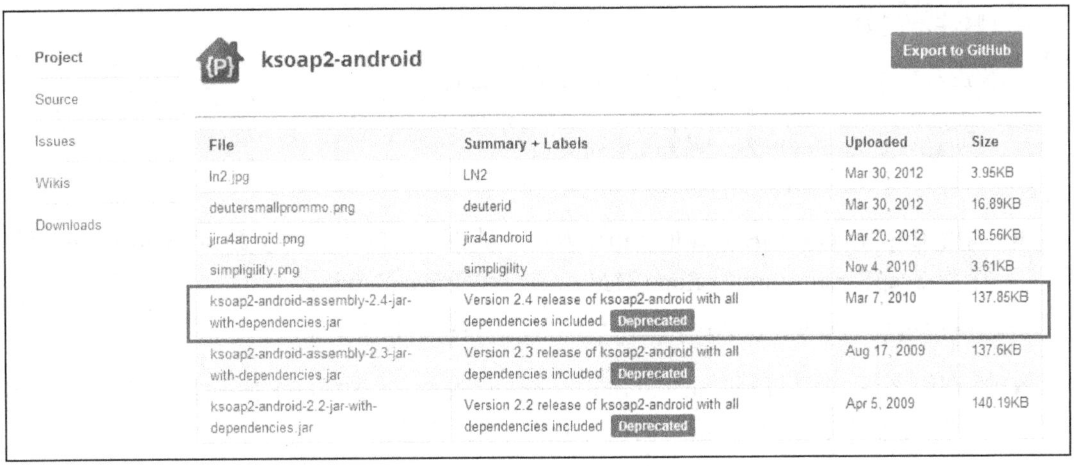

第 10 章　Android 的网络编程基础

图 10.13　下载 ksoap2-android

对于下载有难度的读者，可以直接登录华信教育资源网（www.hxedu.com.cn）找到该 jar 包。

（2）将下载的 ksoap2-android 的 jar 包添加到工程的 lib 目录下，并右键单击选择 "Add as library"，这样就将 ksoap2-android 集成到 Android 项目中了，如图 10.14 所示。

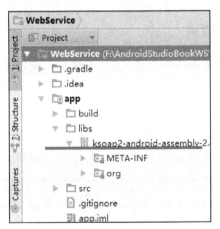

图 10.14　下载 ksoap2-android

使用 kspoap2-android 调用 WebSerice 的操作步骤如下。
- 创建 HttpTransportSE 传输对象，传入 WebService 服务器地址，例如：

```
final HttpTransportSE httpSE = new HttpTransportSE(SERVER_URL);
```

- 创建 SoapObject 对象，创建该对象时需要传入所要调用 WebService 的命名空间、WebService 方法名，例如：

```
SoapObject soapObject = new SoapObject(PACE, M_NAME);
```

- 调用 SoapObject 对象的 **addProperty**(String name,Object value)方法来设置参数，该方法的 name 为参数名；value 为参数值，例如：

```
soapObject.addProperty("byProvinceName", citys);
```

- 创建 SoapSerializationEnelope 对象，并传入 SOAP 协议的版本号；设置对象的 bodyOut

属性，例如：

```
final SoapSerializationEnvelope soapserial = new
            SoapSerializationEnvelope(SoapEnvelope.VER11);
soapserial.bodyOut = soapObject;
soapserial.dotNet = true;
```

- 调用 HttpTransportSE 对象的 call()方法，其中 call 的第一个参数为 soapAction，第二个为 SoapSerializationEvelope 对象，调用远程 WebService：

```
httpSE.call(PACE + M_NAME, soapserial);
```

- 在调用之后，可以访问 SoapSerializationEnvelope 对象的 bodyIn 属性，该属性将返回一个 SoapObject 对象，解析后即可获得结果，例如：

```
SoapObject result = (SoapObject) soapserial.bodyIn;
SoapObject detail = (SoapObject) result.getProperty
                    ("getSupportProvinceResult");
for (int i = 0; i < detail.getPropertyCount(); i++) {
    citys.add(detail.getProperty(i).toString());
}
```

10.4.3 实现手机归属地查询

在了解了 WebService 的基本访问方法之后，就可以根据网上开放的 WebService 服务开发自己的应用程序了。在这一节中将展示如何实现一个手机归属地查询的小应用，我们使用的 WebService 服务地址为：

```
http://ws.webxml.com.cn/WebServices/MobileCodeWS.asmx
```

打开网址后单击"getMobileCodeInfo"进入服务页，如图 10.15 所示。

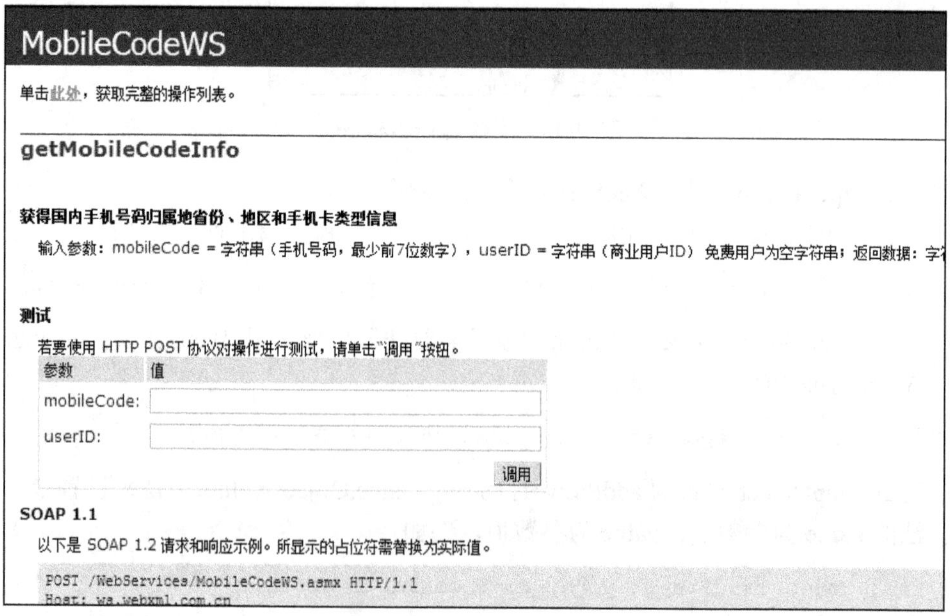

图 10.15　WebService 服务页

第10章 Android 的网络编程基础

在 http://ws.webxml.com.cn/WebServices/MobileCodeWS.asmx 后加上 "?wsdl" 就可以访问其 wsdl 说明，如图 10.16 所示。

```
▼<wsdl:definitions xmlns:soap="http://schemas.xmlsoap.org/wsdl/soap/" xmlns:tm="http://microsoft.com/wsdl/mime/textMatching"
  xmlns:soapenc="http://schemas.xmlsoap.org/soap/encoding/" xmlns:mime="http://schemas.xmlsoap.org/wsdl/mime/" xmlns:tns="http://WebXml.com.cn/"
  xmlns:s="http://www.w3.org/2001/XMLSchema" xmlns:soap12="http://schemas.xmlsoap.org/wsdl/soap12/" xmlns:http="http://schemas.xmlsoap.org/wsdl/http/"
  xmlns:wsdl="http://schemas.xmlsoap.org/wsdl/" targetNamespace="http://WebXml.com.cn/">
 ▼<wsdl:documentation xmlns:wsdl="http://schemas.xmlsoap.org/wsdl/">
    <a href="http://www.webxml.com.cn/" target="_blank">WebXml.com.cn</a> <strong>国内手机号码归属地查询WEB服务</strong>，提供最新的国内手机号码段归属
    用本站 WEB 服务请注明或链接本站:<a href="http://www.webxml.com.cn/" target="_blank">http://www.webxml.com.cn/</a> 感谢大家的支持！<br /> 
  </wsdl:documentation>
 ▼<wsdl:types>
  ▼<s:schema elementFormDefault="qualified" targetNamespace="http://WebXml.com.cn/">
    ▼<s:element name="getMobileCodeInfo">
     ▼<s:complexType>
      ▼<s:sequence>
          <s:element minOccurs="0" maxOccurs="1" name="mobileCode" type="s:string"/>
          <s:element minOccurs="0" maxOccurs="1" name="userID" type="s:string"/>
        </s:sequence>
       </s:complexType>
      </s:element>
    ▼<s:element name="getMobileCodeInfoResponse">
     ▼<s:complexType>
      ▼<s:sequence>
          <s:element minOccurs="0" maxOccurs="1" name="getMobileCodeInfoResult" type="s:string"/>
        </s:sequence>
```

图 10.16　wsdl 说明页

如图 10.16 所示，可以得到以下几个很关键的点。
- 作用域 TargetNameSpace = http://WebXml.com.cn/。
- 查询的方法名为"getMobileCodeInfo"，需要带上 "mobileCode" 与 "userID" 两个参数。
- 返回的结果存在"getMobileCodeInfoResult"中。

【例 10.8】 WebService 实现手机归属地查询实例

代码如下所示：

```java
package com.example.webservice;
import android.os.AsyncTask;
import android.support.v7.app.AppCompatActivity;
import android.os.Bundle;
import android.view.View;
import android.widget.Button;
import android.widget.EditText;
import android.widget.TextView;
import org.ksoap2.SoapEnvelope;
import org.ksoap2.SoapFault;
import org.ksoap2.serialization.SoapObject;
import org.ksoap2.serialization.SoapSerializationEnvelope;
import org.ksoap2.transport.HttpTransportSE;
import org.xmlpull.v1.XmlPullParserException;
import java.io.IOException;

public class WebClient extends AppCompatActivity {
    //URL 地址
    private static final String SERVER_URL =
            "http://ws.webxml.com.cn/WebServices/MobileCodeWS.asmx?wsdl";
    //调用的 WebService 命令空间
    private static final String PACE = "http://WebXml.com.cn/";
    //获取归属地的方法名
    private static final String W_NAME = "getMobileCodeInfo";
```

```java
    private EditText etPhone;
    private Button btnSearch;
    private TextView tvInfo;
    @Override
    protected void onCreate(Bundle savedInstanceState) {
        super.onCreate(savedInstanceState);
        setContentView(R.layout.activity_web_client);
        etPhone = (EditText) findViewById(R.id.etphone);
        btnSearch = (Button) findViewById(R.id.btnsearch);
        tvInfo = (TextView) findViewById(R.id.tvinfo);
        btnSearch.setOnClickListener(new View.OnClickListener() {
            @Override
            public void onClick(View v) {
                String cityName = etPhone.getText().toString();
                if (cityName.length() > 0) {
                    getWeatherInfo(etPhone.getText().toString());
                }
            }
        });
    }
    private void getWeatherInfo(String phoneMum){
        new AsyncTask<String, Void, String>() {
            @Override
            protected String doInBackground(String... params) {
                String local = "";
                final HttpTransportSE httpSe = new
                        HttpTransportSE(SERVER_URL);
                httpSe.debug = true;
                SoapObject soapObject = new SoapObject(PACE, W_NAME);
                //定义soapObject对象
                soapObject.addProperty("mobileCode", params[0]); //输入参数
                soapObject.addProperty("userID", "");
                final SoapSerializationEnvelope serializa = new
                        SoapSerializationEnvelope(SoapEnvelope.VER10);
                //版本选为10
                serializa.setOutputSoapObject(soapObject);
                serializa.dotNet = true;
                //获取返回信息
                try {
                    httpSe.call(PACE + W_NAME, serializa);
                    if (serializa.getResponse() != null) {
                        SoapObject result = (SoapObject) serializa.bodyIn;
                        local = result.getProperty("getMobileCodeInfoResult")
                                .toString();
                    }
                }
                catch (XmlPullParserException e) {
```

第 10 章 Android 的网络编程基础

```
            e.printStackTrace();
        }catch (SoapFault soapFault) {
            soapFault.printStackTrace();
        }catch (IOException e) {
            e.printStackTrace();
        }
        return local;
    }
    @Override
    protected void onPostExecute(String result) {
        tvInfo.setText(result);
    }
}.execute(phoneMum);
    }
}
```

程序运行结果如图 10.17 所示。

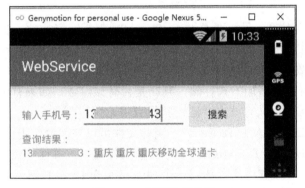

图 10.17 程序运行结果

第 11 章 Android 数据存储

11.1 使用 SharedPreferences

在 Android 中，很多应用都需要存储一些参数，例如在天气 App 中，在这一次使用 App 时添加了多个城市，当用户下一次打开时也希望之前设置的城市会保留在手机中以方便直接获取信息。这个时候就需要 SharedPreferences 类的辅助，利用它来存储一些键值对(key-value)参数。SharedPreferences 类是 Android 提供的一个轻量级的存储类，特别适用于软件的各项参数的存储。

11.1.1 SharedPreferences 的使用方法

首先是获取其实例对象，只需要通过 getSharedPreferences(String, int)就可以直接获取到其实例对象，方法中的第一个参数用于指定该参数的名称，名称不用带后缀，后缀会由 Android 自动加上，第二个参数指定文件的操作模式，共有四种操作模式。
- MODE_APPEND：追加方式存储，判断是否有该参数，如果有则在后面添加。
- MODE_PRIVATE：私有方式存储，其他应用无法访问。
- MODE_WORLD_READABLE：表示当前文件可以被其他应用读取。
- MODE_WORLD_WRITEABLE：表示当前文件可以被其他应用写入。

其次是设置参数，在设置时必须通过一个 SharedPreferences.Editor 对象，以确保在存取时参数保持一致的操作，其操作方法如下所示：

```
Editor editor = sharedPreferences.edit();
```

editor 存储对象采用"key-value"键值对进行存放，例如 editor.putString("name", "hello");这只是对字符串类型的参数进行存放，当然还可以存放更多的类型，方式都为 editor.putXxx("key","value")的形式，Xxx 可以为 Boolean、Float、Int、Long、String。

通过 commit()方法提交了数据，也可以通过 clean()与 remove()方法清除设置的参数。

例如通过下面的代码就可以在安卓的/data/data/<package name>/shared_prefs 目录下生成一个 xml 文件：

```
SharedPreferences sharedPreferences = getSharedPreferences("preference1",
        Context.MODE_PRIVATE);                       //私有数据
Editor editor = sharedPreferences.edit();            //获取编辑器
editor.putString("name", "preference1");
editor.putInt("age", 10);
editor.commit();                                     //提交修改
```

生成的 preference1.xml 文件内容如下：

```
<?xml version='1.0' encoding='utf-8' standalone='yes' ?>
<map>
```

```
<string name="name">preference1</string>
<int name="age" value="10" />
</map>
```

在读取数据时,与之对应的获取数据的方法为 getXxx("key","value"),第二个参数为默认值,如果 preference1 中不存在该 key,将返回该默认值,例如获取上面的参数。

```
SharedPreferences share=getSharedPreferences("preference1",
        Activity.MODE_WORLD_READABLE);
int i=share.getInt("age",0);
```

11.1.2 SharedPreferences 的应用

下面用一个例子说明 SheredPreferences 在记录参数方面的应用。设置用户姓名与年龄并存储,在之后打开 App 时显示存储的数据。

【例 11.1】 使用 SharedPreferences 存储用户信息实例

```java
package com.example.sharedpreference;
import android.app.Activity;
import android.content.SharedPreferences;
import android.os.Bundle;
import android.view.View;
import android.widget.Button;
import android.widget.EditText;
import android.widget.TextView;
public class MainActivity extends Activity {
    EditText edName,edAge;
    TextView tvInfo;
    Button btnCommit;
    private static final String FILENAME = "info";        //定义文件名
    @Override
    protected void onCreate(Bundle savedInstanceState) {
        super.onCreate(savedInstanceState);
        setContentView(R.layout.activity_main);
        edName = (EditText)findViewById(R.id.edname);
        edAge = (EditText)findViewById(R.id.edage);
        btnCommit = (Button)findViewById(R.id.btncommit);
        tvInfo = (TextView)findViewById(R.id.tvinfo);
        SharedPreferences preferences = super.getSharedPreferences
            (FILENAME, Activity.MODE_PRIVATE);            //获取文件
        tvInfo.setText("用户信息 姓名: " + preferences.getString("name",
           "未定义") + ", 年龄: " + preferences.getInt("age", 0) + "。");
                                                          //初始化内容
        btnCommit.setOnClickListener(new View.OnClickListener() {
            @Override
            public void onClick(View v) {
                SharedPreferences preferences = getSharedPreferences
```

```
                (FILENAME, Activity.MODE_PRIVATE);  //获取文件
                SharedPreferences.Editor edit = preferences.edit();
                edit.putString("name", edName.getText().toString());
                edit.putInt("age", Integer.parseInt(edAge.getText().toString()));
                edit.commit();     //提交修改
                tvInfo.setText("用户信息 姓名："+edName.getText().toString()
                        +"，年龄："+edAge.getText().toString());
            }
        });
    }
}
```

在设置姓名为"admin"、年龄为 12 之后，关闭程序，再次打开的运行效果如图 11.1 所示。"info"文件存放在哪里呢？在 Android Studio 中打开 Android Device Monitor，如图 11.2 所示。

图 11.1　程序运行效果

图 11.2　打开 Android Device Monitor

然后选择"File Explorer"进入到 data\data\<packagename>\shared_prefs，就可以看到 info.xml 文件，如图 11.3 所示。

图 11.3　文件存储路径

选中文件，单击窗口右上角的"导出文件"图标就可以导出文件到计算机上查看，如图 11.4 所示。

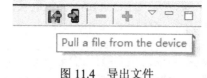
图 11.4　导出文件

在计算机查看 xml 文件内容如下：

```
<?xml version="1.0" encoding="UTF-8" standalone="true"?>
-<map>
<int value="12" name="age"/>
<string name="name">tt</string>
</map>
```

11.2 File 存储

11.2.1 使用 IO 流操作文件

Java 中提供的 IO 流操作在 Android 中也同样适用，context 所提供的 openFileOutput 与 openFileInput 就能十分方便地操作文件，具体实现过程如下所示。

(1) 文件的存储：

```
FileOutputStream out = context.openFileOutput(String filename,int mode);
                                        //以 mode 模式获得文件输出流
out.write(byte[]);                      //写入内容
```

(2) 文件的读取：

```
FileInputStream in = context.openFileInput(String filename);
                                        //获得某个文件的文件流
int length = in.read(byte[]);           //读取内容
```

(3) 文件操作模式与之前提到的 SharedPreferences 类中的模式相似，也有四种操作模式。

- Context.MODE_PRIVATE：私有覆盖模式。
  只能被当前应用访问，并且如果写入，则覆盖。
- Context.MODE_APPEND：私有追加模式。
  只能被当前应用访问，并且如果写入，则追加。
- Context,MODE_WORLD_READABLE：公有只读模式。
  可以被其他应用读取。
- Context.MODE_WORLD_WRITEABLE：公有可写模式。
  可以被其他应用写入，但不能读取。

11.2.2 文件操作应用

下面用一个例子说明 IO 流在文件操作中的应用。将以特定的文件名称和特定的文件内容进行保存，单击读取时，将读取特定的文件的文件内容显示到文件内容文本框。

【例 11.2】 File 文件存取实例

```
package com.example.file;
import android.app.Activity;
import android.content.Context;
import android.os.Bundle;
import android.view.View;
import android.widget.Button;
import android.widget.EditText;
import java.io.ByteArrayOutputStream;
import java.io.FileInputStream;
import java.io.FileOutputStream;
public class MainActivity extends Activity {
    private Button btnSave,btnRead;
```

```java
    private EditText edFilename,edFilecontent;
    private Context context = this;
    @Override
    public void onCreate(Bundle savedInstanceState) {
        super.onCreate(savedInstanceState);
        setContentView(R.layout.activity_main);
        btnSave = (Button)this.findViewById(R.id.btnSave);
        btnRead = (Button)this.findViewById(R.id.btnRead);
        edFilename = (EditText)this.findViewById(R.id.edFilename);
        edFilecontent = (EditText)this.findViewById(R.id.edFilecontent);
        btnSave.setOnClickListener(new View.OnClickListener(){
            @Override
            public void onClick(View v) {
                String filename = edFilename.getText().toString();
                String filecontent = edFilecontent.getText().toString();
                FileOutputStream out = null;
                try {
                    out = context.openFileOutput(filename, Context
                            .MODE_PRIVATE);
                    out.write(filecontent.getBytes("UTF-8"));
                } catch (Exception e) {
                    e.printStackTrace();
                }
                finally{
                    try {
                        out.close();
                    } catch (Exception e) {
                        e.printStackTrace();
                    }
                }
            }
        });

        btnRead.setOnClickListener(new View.OnClickListener(){
            @Override
            public void onClick(View v) {
                String filename = edFilename.getText().toString();
                                                            //获得文件的名称
                FileInputStream in = null;
                ByteArrayOutputStream bout = null;
                byte[]buf = new byte[1024];
                bout = new ByteArrayOutputStream();
                int length = 0;
                try {
                    in = context.openFileInput(filename);   //获得输入流
                    while((length=in.read(buf))!=-1){
                        bout.write(buf,0,length);
                    }
```

第 11 章 Android 数据存储

```
            byte[] content = bout.toByteArray();
            edFilecontent.setText(new String(content,"UTF-8"));
                                            //设置文本框为读取的内容
        } catch (Exception e) {
            e.printStackTrace();
        }
        edFilecontent.invalidate();    //刷新屏幕
        try{
            in.close();
            bout.close();
        }
        catch(Exception e){}
    }});
}}
```

对布局文件不再赘述，读者可以自行编写，实现效果如图 11.5 所示。

图 11.5 文件读取应用

文件存储位置如图 11.6 所示。

图 11.6 文件存储位置

导出后即可查看文件内容与输入一致，如图 11.7 所示。

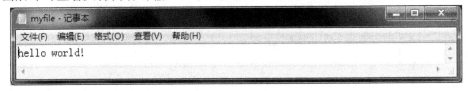

图 11.7 文件内容

11.2.3 将文件保存到 SD 卡

在上述的操作中，直接将文件保存在了默认的存储空间中。如果一个文件很大，则通常存放在手机的 SD 卡中；如果手机存在 sdcard，则 sdcard 的目录为 **/mnt/sdcard** 目录。

在 Genymotion 中下载安装虚拟机的同时,便会指定存储空间的大小,如图 11.8 所示。

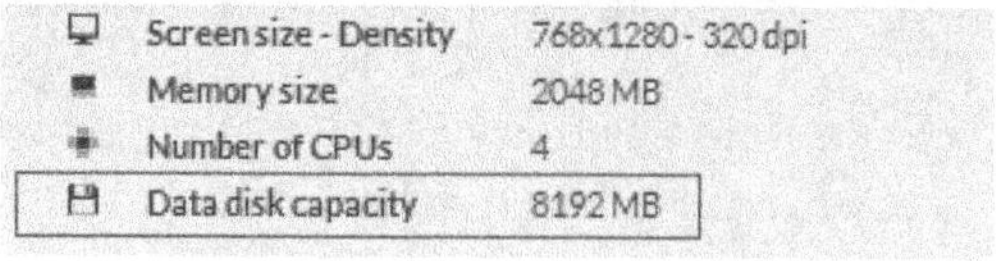

```
Screen size - Density    768x1280 - 320 dpi
Memory size              2048 MB
Number of CPUs           4
Data disk capacity       8192 MB
```

图 11.8　手机内部存储

在 Android 中读取 SD 卡时,需在 AndroidManifest.xml 中设置相关权限。

```
<uses-permission android:name=
"android.permission.MOUNT_UNMOUNT_FILESYSTEMS"></uses-permission>
<uses-permission android:name=
"android.permission.WRITE_EXTERNAL_STORAGE"></uses-permission>
```

在 Android 中,1.5、1.6 版本的 sdcard 目录为/sdcard,而 Android 2.0 以上都是/mnt/sdcard。因此,如果在保存时直接写具体目录就不行了,可以使用 Environment 类去获取外部存储目录,例如这个类提供的 getExternalStorageDirectory()方法就能直接获取 sdcard 目录。还有一种情况是不确定一部手机是否存在 sdcard,这时也需要借助 Environment 类进行判断。如果没有,则需要提供其他解决方法,比如保存到手机存储,提示不存在 sdcard,方法如下:

```
if(Environment.getExternalStorageState()
    .equals(Environment.MEDIA_MOUNTED)){
    //执行存储 sdcard 方法  }
else{  //存储到手机中,或提示无 SD 卡  }
```

除上述的方法外,Environment 类还提供了一些其他方法,可让用户更方便地操作外部存储目录。Environment 类的常量与常用方法如表 11.1 所示。

表 11.1　Environment 类的常量与常用方法

返回类型	常量或方法名	描述
	MEDIA_BAD_REMOVAL	存储设备被移除
	MEDIA_CHECKING	存储设备正在检查
	MEDIA_MOUNTED	存储设备可读/写
	MEDIA_MOUNTED_READ_ONLY	存储设备只读
	MEDIA_REMOVED	存储设备不存在
static File	getDataDirectory()	获取用户 data 目录
static File	getDownloadCacheDirectory()	获取下载内容目录
static File	getExternalStorageDirectory()	获取扩展存储目录
static String	getExternalStorageState(File path)	获取扩展存储状态
static File	getRootDirectory()	获取根目录

接着上面的例子,可以再添加两个按钮,将数据保存在 SD 卡中,以及从 SD 卡中读取存储的数据,两个按钮的监听代码如下。

写入到 SD 卡:

```
btnSaveSD.setOnClickListener(new View.OnClickListener(){
    @Override
    public void onClick(View v) {
```

```
            String filename = edFilename.getText().toString();
            String filecontent = edFilecontent.getText().toString();
if(Environment.getExternalStorageState().equals(Environment.MEDIA_MOUNTED)){
    File file = new File(Environment.getExternalStorageDirectory().toString()
        + File.separator+"cqupt"+File.separator + filename);  //定义存储路径
            if(!file.getParentFile().exists()){
                file.getParentFile().mkdirs();        //创建文件夹
            }
            PrintStream out = null;
            try{
                out = new PrintStream(new FileOutputStream(file,true));
                out.println(filecontent);           //写入内容
            }catch (Exception e){
                e.printStackTrace();
            }
            finally {
                if (out!= null){
                    out.close();
                }
            }
        }
        else {
            Toast.makeText(MainActivity.this, "SD卡不存在，存储失败",
                Toast.LENGTH_SHORT).show();
        }
    }});
```

从 SD 卡中读取：

```
btnReadSD.setOnClickListener(new View.OnClickListener() {
    @Override
    public void onClick(View v) {
        String filename = edFilename.getText().toString();
                                                //获得读取的文件的名称
        String filecontent = "";
 if(Environment.getExternalStorageState().equals(Environment
            .MEDIA_MOUNTED)){
 File file = new File(Environment.getExternalStorageDirectory().toString()
        + File.separator + "cqupt" + File.separator + filename);
            if (!file.getParentFile().exists()){
                file.getParentFile().mkdirs();
            }
            Scanner scan = null;
            try{
                scan = new Scanner(new FileInputStream(file));
                while(scan.hasNext()){
                    filecontent += scan.next()+"\n";
                }
                edFilecontent.setText(filecontent);
```

```
                }catch (Exception e){
                    e.printStackTrace();
                }finally {
                    if (scan!=null){
                        scan.close();
                    }
                }
            }else {
                Toast.makeText(MainActivity.this, "SD 卡不存在，存储失败",
                        Toast.LENGTH_SHORT).show();
            }
        }
    });
```

这也同样实现了文件的读取操作，只不过把数据放在了外部存储中。

11.3 SQLite 数据库

通过以上介绍，已经能够存储简单的键值对或者其他类型的文件，但在软件项目中，这些是远远不够的。有时，需要对数据表单进行处理，一些数据表甚至达到上百上千行。如果没有一个稳健的数据库存储机制，是无法进行这样的工作的。所以，Android 中也支持 SQLite 数据库，用其处理复杂的数据表。SQLite 是轻量级数据存储工具，是遵守 ACID（增删改查）的关系型数据库管理系统。它具备 SQL 数据库的所有优点，能够通过 SQL 语句操作数据表。

11.3.1 SQLite 介绍

在 Android 上使用 SQLite 之前，需要先了解 SQLite 的存储机制以及 SQL 常用操作语句。SQLite 是一个开源的嵌入式关系数据库，实现自包容、零配置、支持事务的 SQL 数据库引擎。SQLite 在易用性、便携性、紧凑性、高效性和可靠性方面的表现十分优秀，SQLite 的安装和运行非常简单，在大多数情况下只要确保 SQLite 的二进制文件存在即可开始创建、连接和使用数据库。如果需要一个嵌入式数据库项目或解决方案，SQLite 是绝对值得考虑的。

SQLite 具有一个十分简洁明了的体系结构，如图 11.9 所示。

如图 11.9 所示，这些模块的分工清晰，将查询过程划分为几个独立的任务，就像在流水线上操作一样。首先让开发人员通过开放的接口与 SQLite 进行交互；然后再经过 Tokenizer（分析器）与 Parser（语法分析器）进行编译，转化为底层能更方便处理的层次化数据结构；之后再交给虚拟机完成具体的数据库操作（比如打开一个表的游标、查询一列等）；最后在后端中负责数据的排序、读/写等操作。

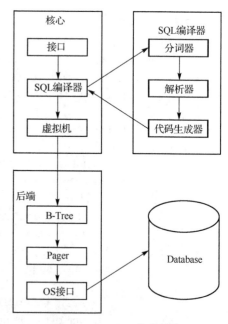

图 11.9　SQLite 体系结构

在了解 SQLite 的机制之后，还需要学习 SQLite 中的 SQL。SQL 是学习数据库的基础，也是关系数据库通信的唯一渠道。使用 SQL 能够十分便捷地进行构建、读/写、排序、过滤、计算、分组等常用的信息管理。本节主要介绍 SQLite 中 SQL 语句的使用，不做深入讲解。

1) 语法

大体上可以把 SQL 分为两个部分：数据操作语言(DML)和数据定义语言(DDL)。

查询和更新指令构成了 DML 部分：

- select：从数据库表中获取数据。
- update：更新数据库表中的数据。
- delete：从数据库表中删除数据。
- insert into：向数据库表中插入数据。

SQL 的数据定义语言(DDL)部分使我们有能力创建或删除表格。也可以定义索引(键)，规定表之间的链接，以及施加表间的约束。

SQL 中最重要的 DDL 语句如下。

- create database：创建新数据库。
- alter database：修改数据库。
- create table：创建新表。
- alter table：变更(改变)数据库表。
- drop table：删除表。
- create index：创建索引(搜索键)。
- drop index：删除索引。

SQL 采用声明式的语法规则，就像自然语言的表达方式一样，可以清楚地描述需要，而不需要制定如何做。下面以处理一个成绩表单为例，该表如表 11.2 所示。

表 11.2 "Grades" 表

id	name	sex	chinese	math
1	Xiaoming	male	89	99
2	Xiaoming	female	78	89
3	Zhangwei	male	85	92

在查询男生小明的语文成绩时，使用 SQL 语句的表达方式为：

SELEDT id FROM Grades WHERE name='Xiaoming' AND sex='male';

（谓语）（主语）（宾语）

正如上述语句所示，SQL 非常简单，交互方式十分便捷。但要注意，上述语句中有指定含义的单词(如"SELECT"、"UPDATE"、"DELETE"、"INSERT"、"DROP"等)都是关键字，SQL 不区分关键字的大小写，例如下面的写法都是相同的语句：

```
SELECT name from Grades;
seLeCt name From Grades;
```

2) 创建数据表

在上述内容中可以知道，创建表的语句属于 DDL 部分。可以用 CREATE TABLE 创建表，在创建表时需要指定表名、字段名、字段类型以及其约束。在 SQLite 中有以下几种数据类型。

- varchar(n)：长度不固定且其最大长度为 n 的字串，n 不能超过 4000。
- char(n)：长度固定为 n 的字串，n 不能超过 254。
- integer：值被标识为整数。
- real：所有值都是浮动的数值，被存储为 8 字节的 IEEE 浮动标记序号。
- text：值为文本字符串，使用数据库编码存储。
- blob：值是 blob 数据块，以输入的数据格式进行存储。如何输入就如何存储，不改变格式。
- data：包含年份、月份、日期。
- time：包含小时、分钟、秒。

可以看到这几种数据类型对于学过数据库的来说并不陌生。"约束"用来控制什么样的值可以存储在该字段中。如果需要创建如表 11.2 所示的表，其 SQL 语句如下：

```
create table grades(
    id        integer   identity(1,1)  primary key,
    name      varchar   not            null,
    sex       varchar   not            null,
    chinese   integer   not            null,
    math      integer   not            null
)
```

在上述语句中，列属性"identity(起始值，递增量)"表示"id"列为自动编号，也称为标识列。

3) 数据库常用操作

- 查询：select 作为查询数据库的唯一命令，是 SQL 中功能最强大的但也是最复杂的命令。select 命令采用一系列的字句将很多关系操作组合在一起。从而可以在数据表中只选择获取所需要的数据。

select 命令的通用形式如下：

```
select [distinct] heading
    from tables                    //指定在哪一个表中
    where predicate                //对字段进行过滤
    group by columns               //分组
    having predicate
    order by columns               //按照什么顺序
    limit count,offset;            //限定与排序
```

- 增加：使用 insert 命令向表中插入记录。使用 insert 命令可以一次插入一条记录，insert 语句的一般格式为：

```
insert into table_name (list1,list2…) values (value1,value2…);
```

变量 table_name 表示数据插入到哪个表中，list1，list2…是用逗号分隔的字段名称。这些字段必须是表中存在的，并一一对应于 value1,value2…。

- 更新：使用 update 命令用于更新表中的记录，使用 update 命令可以修改一个表中的一行或者多行中的一个或多个字段，update 语句的一般格式为：

```
update 表名称 set 列名称 = 新值 where 列名称 = 某值;
```

- **删除**：使用 delete 命令可以删除表中的记录，delete 语句的一般格式为：

```
delete from 表名称 where 列名称 = 值;
```

11.3.2 SQLite 数据库操作

Android 平台为了方便开发者操作 SQLite，封装了创建和使用 SQLite 数据库的 API，从而可以在应用程序中管理自有的数据库。Android 中提供的 SQLite 操作类如表 11.3 所示。

表 11.3　Android 中提供的 SQLite 操作类

类名	描述
SQLiteCursor	游标，实现从数据库中返回查询结果
SQLiteDatabase	提供管理 SQLite 数据库的操作方法
SQLiteOpenHelper	辅助类来管理数据库的建立和版本
SQLiteQuery	读取结果行

每一个 SQLiteDatabase 类都代表一个数据库对象，提供了操作数据库的一些方法。在 Android 的 SDK 目录下有 sqlite3 工具，可以利用它创建数据库、创建表和执行一些 SQL 语句。SQLiteDatabase 类的常用方法如表 11.4 所示。

表 11.4　SQLiteDatabase 类的常用方法

返回类型	方法名	描述
static SQLiteDatabase	openDatabase(String path, SQLiteDatabase.CursorFactory factory, int flags)	打开指定路径下的数据库
static SQLiteDatabase	openOrCreateDatabase(String path, SQLiteDatabase.CursorFactory factory)	打开或创建指定路径下的数据库
long	insert(String table,String nullColumnHack,ContentValues values)	插入一行到数据库
int	delete(String table,String whereClause,String[] whereArgs)	删除数据库某一行
int	update(String table,ContentValues values,String whereClause,String[] whereArgs)	更新数据库某一行
Cursor	query(String table,String[] columns,String selection,String[] selectionArgs,String groupBy,String having,String orderBy)	数据表查询
Cursor	rawQuery(String sql, String[] selectionArgs)	使用 SQL 语句进行数据表查询
void	execSQL(String sql)	执行 SQL 语句
void	setVersion(int version)	设置数据库版本
void	close()	关闭数据库

SQLiteDatabase 类本身只是一个数据库的操作类，但是如果需要对数据库进行操作，则还要得到 SQLiteOpenHelper（数据库操作辅助类）的帮助。这个类主要生成一个数据库，并对数据库的版本进行管理。SQLiteOpenHelper 是一个抽象类，通常需要继承它，并且实现以下 3 个函数。

- onCreate(SQLiteDatabase)：在数据库第一次生成的时候会调用这个方法，也就是说，只有在创建数据库的时候才会调用。当然，也有一些其他的情况，一般在这个方法里生成数据库表。

- onUpgrade（SQLiteDatabase, int, int）：当数据库需要升级的时候，Android 系统会主动调用这个方法。一般在这个方法里删除数据表，并建立新的数据表。当然，是否还需要做其他的操作完全取决于应用的需求。
- onOpen（SQLiteDatabase）：这是当打开数据库时的回调函数，一般在程序中不常使用。

这个类也提供了一些其他常用的数据库操作方法，如表 11.5 所示。

表 11.5 SQLiteOpenHelper 类的常用方法

返回类型	方法名	描述
String	getDatabaseName()	获取数据库名称
SQLiteDatabase	getReadableDatabase()	以只读方式创建或打开数据库
SQLiteDatabase	getWritableDatabase()	以修改方式创建或打开数据库
void	onOpen(SQLiteDatabase db)	当数据库已经打开时调用
int	update(String table,ContentValues values,String whereClause,String[] whereArgs)	更新数据库某一行
void	close()	关闭数据库

在了解了 SQLite 的常用操作方式之后，便可以开始建立数据库进行实践练习了。下面通过学习一个实例，展示如何使用 SQLite。

1．创建数据库

定义 SQLiteOpenHelper 的子类，在该类中创建并打开一个指定名称的数据库对象。

```
package com.example.sqlite;
import android.content.Context;
import android.database.sqlite.SQLiteDatabase;
import android.database.sqlite.SQLiteOpenHelper;
public class MyDatabaseHelper extends SQLiteOpenHelper {
    private static final String DATABASE_NAME = "sqlite.db";
    private static final int DATABASE_VERSION = 1;
    private static final String TABLE_NAME = "grades";
    //调用父类构造器
    public MySQLiteHelper(Context context, String name, SQLiteDatabase
.CursorFactory factory, int version) {
        super(context, DATABASE_NAME, factory, DATABASE_VERSION);
    }
    /**
     * 当数据库首次创建时执行该方法，一般将创建表等初始化操作放在该方法中
     * 执行，重写 onCreate 方法，调用 execSQL 方法创建表
     * */
    @Override
    public void onCreate(SQLiteDatabase db) {
        db.execSQL("create table" + "if not exists" + TABLE_NAME + "("
            + "id integer primary key,"
            + "name varchar,"
            + "sex varchar,"
            + "chinese integer,"
            + "math integer)");
```

```
        }
    //当打开数据库时传入的版本号与当前的版本号不同时会调用该方法
    @Override
    public void onUpgrade(SQLiteDatabase db, int oldVersion, int newVersion) {
        db.execSQL("drop table if exists" + TABLE_NAME);
        this.onCreate(db);              //创建表
    }
}
```

然后定义 MainActivity.java,调用 MyDatabaseHelper 类完成表的创建。

```
package com.example.sqlite;
import android.support.v7.app.AppCompatActivity;
import android.os.Bundle;
public class MainActivity extends AppCompatActivity {
MySQLiteHelper helper;
public static final String TABLE_NAME = "grades";
    @Override
    protected void onCreate(Bundle savedInstanceState) {
        super.onCreate(savedInstanceState);
        setContentView(R.layout.activity_main);
        helper = new MySQLiteHelper(this);
        helper.getWritableDatabase();
    }
}
```

通过以上代码轻松地建立了一个数据库,并在数据库中插入了"grades"数据表。在 Genymotion 模拟器上运行程序之后,我们就可以同 11.1 节中查看存储文件同样的方式获取 db 文件,db 文件的存放路径为 data/data/com.example.sqlite/databases/sqlite.db,如图 11.10 所示。

图 11.10 文件存储路径

如果计算机上没有安装相应的软件,则无法打开 db 文件。在安装了 SQLite Expert 软件后就可以查看数据表,如图 11.11 所示。

图 11.11 查看 db 文件

2. 修改数据表中的数据

在创建了数据库之后，可以直接使用 SQL 语句操作数据库。在主程序中添加如下数据库修改方法即可。

```
public void insert(String name, String sex, int chinese, int math){
    SQLiteDatabase db = helper.getWritableDatabase();
    db.execSQL("insert into "+ TABLE_NAME +"(name,sex,chinese,math)
        values('"+ name +"','" + sex + "','"+ chinese+"','"+ math +"')");
    db.close();
}

public void update(int id, String name, String sex, int chinese, int math){
    SQLiteDatabase db = helper.getWritableDatabase();
    db.execSQL("update "+ TABLE_NAME +" set name ='" + name + "',sex='" +
        sex + "',chinese='" + chinese + "',math='" + math +"'where id=" + id);
    db.close();
}
public void delete(int id)
{
    SQLiteDatabase db = helper.getWritableDatabase();
    db.execSQL("delete from " +TABLE_NAME+" where id=" + id);
    db.close();
}
```

如果在表中依次调用这些函数则可以看到数据库的内容变化。

执行插入语句：

```
insert("Xiaoming","male", 89, 99);
insert("Xiaoming", "female", 78, 89);
insert("Zhangwei","male",85,92);
update(1, "Xiaohong", "female", 85, 97);
delete(3);
```

结果如图 11.12 所示。

RecNo	id	name	sex	chinese	math
		Click here to define a filter			
1	1	Xiaoming	male	89	99
2	2	Xiaoming	female	78	89
3	3	Zhangwei	male	85	92

RecNo	id	name	sex	chinese	math
		Click here to define a filter			
1	1	Xiaoming	male	89	99
2	2	Xiaohong	female	85	97
3	3	Zhangwei	male	85	92

RecNo	id	name	sex	chinese	math
		Click here to define a filter			
1	1	Xiaoming	male	89	99
2	2	Xiaohong	female	85	97

图 11.12　数据库修改结果

在上述方法中直接使用 SQL 语句对数据表进行操作，可以看到语句中的符号很多，特别容易出错，所以为了更方便地操作数据，SQLiteDatabase 类中也提供了 insert()、update()、delete() 等数据操作方法。另外，在这些方法中还可以使用 ContentValues 封装数据。ContentValues 的使用方式类似于 SharedPreference 类中提供的键值对数据存储，表 11.6 展示了 ContentValues 类提供的常用方法：

表 11.6 ContentValues 类提供的常用方法

返回类型	方法名	描述
void	clear()	清空数据
boolean	containsKey(String key)	判断是否包含该字段
void	put(String key, Xxx value)	添加值到指定的字段中，Xxx 为字段类型
Xxx	getAsXxx(String key)	获取字段中的值，Xxx 为字段类型
int	size()	返回数值

所以可以将上述的各个方法修改为：

```java
public void insert(String name, String sex, int chinese, int math){
    SQLiteDatabase db = helper.getWritableDatabase();
    ContentValues values = new ContentValues();
    values.put("name", name);
    values.put("sex", sex);
    values.put("chinese",chinese);
    values.put("math",math);
    db.insert(TABLE_NAME, "id", values);
    db.close();
}

public void update(int id, String name, String sex, int chinese, int math){
    SQLiteDatabase db = helper.getWritableDatabase();
    ContentValues values = new ContentValues();
    values.put("name", name);
    values.put("sex", sex);
    values.put("chinese",chinese);
    values.put("math",math);
    db.update(TABLE_NAME, values,"id = "+id,null);
    db.close();
}

public void delete(int id)
{
    SQLiteDatabase db = helper.getWritableDatabase();
    db.delete(TABLE_NAME,"id = "+id,null);
    db.close();
}
```

可以看出这样的操作比较简单明了。

3．查询表中的数据

在建立好数据表之后，下一步是如何查询需要的信息。SQL 提供了 select 这个功能强大而又复杂的查询语句。同上述的操作方式一样，可以使用 SQLiteDatabase 类所提供的数据库操作方法，如 rawQuery()或者 Query()方法。在学习这些查询方法之前，需要了解 Cursor 这个类。Cursor 就是游标，类似于在数据表中游走，用于对结果集进行随机访问，Cursor 与 JDBC 中的 ResultSet 作用很相似。Cursor 类也提供了一些常用方法，如表 11.7 所示。

表 11.7 Cursor 类的常用方法

返回类型	方法名	描述
void	close()	关闭游标，释放数据库资源
int	getCount()	返回数据库行数
int	getPosition()	返回游标所在的行数
int	getColumnIndex(String columnName)	返回查询字段的列，当该值不存在时返回-1
String	getString(int columnIndex)	以字符串的形式返回该列的值
boolean	moveToNext()	移动游标到下一行
boolean	MoveToPrevious()	移动游标到上一行
boolean	moveToFirst()	移动游标到第一行
boolean	moveToLast()	移动游标到最后一行

同之前的数据库操作方式一样，查询既可以使用 SQL 语句也可使用 SQLiteDatabase 类所提供的 query()方法。

假设现在需要查询某一位同学的语文成绩，可以通过 rawQuery()方法实现带参数查询：

```java
public int getChinese(String name){
    int grade = -1;
    Cursor c = db.rawQuery("select * from " +TABLE_NAME+" where name=?",name);
    if(cursor.moveToFirst()) {
        grade = c.getString(c.getColumnIndex("chinese"));   //获取相关的分数
    }
    return grade;
}
```

或者使用 query()方法，完整带参数的方法如下所示：

```
query(String table, String[] columns, String selection, String[] selectionArgs, String groupBy, String having, String orderBy, String limit)
```

其中各个字段的含义如下。
- table：数据表名称。
- columns：列名称数组。
- selection：条件子句，相当于 where。
- selectionArgs：条件语句的参数数组。
- groupBy：分组。
- having：分组条件。
- orderBy：排序方式，可以选定一字段以升序或者降序的方式排列。
- limit：分页查询的限制。

上述的查询方法可改为：

```java
public int getChinese(){
    Cursor c = db.query(TABLE_NAME,null,null,null,null,null,null);
```

```
        if(c.moveToFirst()){                //查询并获得游标
                                             //判断游标是否为空
          grade = c.getString(c.getColumnIndex("chinese"));
        }
}
```

学会了查询语句的使用方式之后,可以完成一个完整的数据库操作小程序。下面完成一个简易的记账软件开发,在该软件中可以定义金额的数值、金额的类型(支出或收入)和描述,并且可以查询当前总的支出与收入。

【例 11.3】 记账小软件实例

建立 DBHelper 类,该类继承于 SQLiteOpenHelper,用于数据库创建,DBHelper.java 代码如下所示:

```java
package com.example.billbook;
import android.content.Context;
import android.database.sqlite.SQLiteDatabase;
import android.database.sqlite.SQLiteOpenHelper;
public class DBHelper extends SQLiteOpenHelper {
    private static final String DATABASE_NAME = "sqlite.db";  //数据库名称
    private static final String TABLE_NAME = "count";          //表名称
    private static final int VERSION = 1;                       //数据库版本
    public DBHelper(Context context) {
        super(context, DATABASE_NAME, null, VERSION);
    }
    @Override
    public void onCreate(SQLiteDatabase db) {
        db.execSQL("create table " + "if not exists " + TABLE_NAME + " ("
            + "id integer primary key,"
            + "count varchar,"
            + "type varchar,"
            + "date varchar,"
            + "describe varchar)");
    }
    @Override
    public void onUpgrade(SQLiteDatabase db, int oldVersion, int newVersion) {
        db.execSQL("drop table if exists" + TABLE_NAME);
        this.onCreate(db);
    }
}
```

接着定义 Count 类用于封装账单数据,Count.java 代码如下:

```java
package com.example.billbook;
public class Count {
    private Double money;
    private String type;
    private String date;
    private String describe;
    public Count() {
    }
```

```java
public Count(Double money, String date, String type, String describe) {
    this.money = money;
    this.date = date;
    this.type = type;
    this.describe = describe;
}
public Double getMoney() { return money; }
public void setMoney(Double money) { this.money = money; }
public String getType() { return type; }
public void setType(String type) { this.type = type; }
public String getDate() { return date; }
public void setDate(String date) {this.date = date; }
public String getDescribe() { return describe; }
public void setDescribe(String describe) { this.describe = describe; }
}
```

接着定义 DBManager 类，用于维护管理数据库，包括数据插入、数据的查询等操作，DBManager.java 代码如下：

```java
package com.example.billbook;
import android.content.ContentValues;
import android.content.Context;
import android.database.Cursor;
import android.database.sqlite.SQLiteDatabase;
public class DBManager {
    private DBHelper helper;
    private SQLiteDatabase db;
    private static final String TABLE_NAME = "count";
    public DBManager(Context context) {
        helper = new DBHelper(context);
        db = helper.getWritableDatabase();
    }
    public void insert(Count count) {
        db.beginTransaction();                  //开始事务
        try {
            ContentValues cv = new ContentValues();
            cv.put("count",count.getMoney());
            cv.put("type",count.getType());
            cv.put("date", count.getDate());
            cv.put("describe",count.getDescribe());
            db.insert(TABLE_NAME,"id",cv);
            db.setTransactionSuccessful();      //设置事务成功完成
        }finally {
            db.endTransaction();                //结束事务
        }
    }
    public Double getResult(int type)
    {
```

```
            Double result = 0.0;
            Cursor c = db.rawQuery("select id,count,type,date
                ,describe from "+ TABLE_NAME,null);
            for (c.moveToFirst();!c.isAfterLast();c.moveToNext()) {
                if (c.getInt(2) == type)
                    result += Double.parseDouble(c.getString(1));
            }
            c.close();
            return result;
        }
        public void closeDB(){
            db.close();}
}
```

最后在 MainActivity.java 中实例化数据库,并对 UI 空间添加监听,完成记账功能。代码如下:

```
package com.example.billbook;
import android.support.v7.app.AppCompatActivity;
import android.os.Bundle;
import android.view.View;
import android.widget.Button;
import android.widget.EditText;
import android.widget.RadioGroup;
import android.widget.TextView;
import java.text.SimpleDateFormat;
import java.util.Date;

public class MainActivity extends AppCompatActivity {
    private static final int OUT = 1;
    private static final int IN = 2;
    private DBManager mgr;
    private EditText etCount,etDescribe;
    private RadioGroup radioGroup;
    private TextView tvInfo;
    private Button btnAdd;
    private int countType = OUT;
    @Override
    protected void onCreate(Bundle savedInstanceState) {
        super.onCreate(savedInstanceState);
        setContentView(R.layout.activity_main);
        //初始化DBManager
        mgr = new DBManager(this);
        //获取UI控件
        etCount = (EditText) findViewById(R.id.etCount);
        etDescribe = (EditText) findViewById(R.id.etDescribe);
        radioGroup = (RadioGroup) findViewById(R.id.radioGroup);
        tvInfo = (TextView)findViewById(R.id.tvInfo);
```

```java
        btnAdd = (Button)findViewById(R.id.btnAdd);
        radioGroup.setOnCheckedChangeListener(new
                    RadioGroup.OnCheckedChangeListener() {
            @Override
            public void onCheckedChanged(RadioGroup group, int checkedId) {
                switch (checkedId) {
                    case R.id.radioOut:
                        countType = OUT;//定义为支出
                        break;
                    case R.id.radioIn:
                        countType = IN; //定义为收入
                        break;
                }
        }});
        btnAdd.setOnClickListener(new View.OnClickListener() {
            @Override
            public void onClick(View v) {
                Count count = new Count();
                long time = System.currentTimeMillis();    //获取系统时间
                SimpleDateFormat format = new SimpleDateFormat("yyyy-MM-dd
                    HH:mm:ss");                            //格式化时间
                String str = format.format(new Date(time));  //转化为字符串
                count.setDate(str);
                count.setMoney(Double.parseDouble(etCount.getText()
                    .toString()));
                count.setDescribe(etDescribe.getText().toString());
                count.setType(countType + "");
                mgr.insert(count);                          //插入数值
                resetInfo();                                //刷新结果
            }
        });
        resetInfo();
    }
    public void resetInfo(){
        Double out = mgr.getResult(OUT);
        Double in = mgr.getResult(IN);
        Double all = in - out;
        tvInfo.setText("总计支出："+out+"   总计收入："+in+"结余："+all+"。");
    }
    protected void onDestroy() {
        super.onDestroy();
        //应用的最后一个Activity关闭时应释放DB
        mgr.closeDB();
    }
}
```

可以看到显示效果如图 11.13 所示。

第 11 章 Android 数据存储

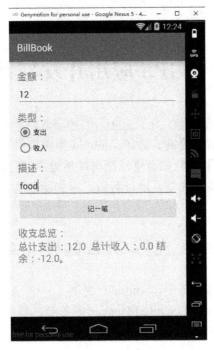

图 11.13 软件显示效果

第 12 章　GPS 应用开发与百度地图

随着随着智能手机的逐渐普及，为了适应人们的需求，手机定位已经成为必不可少的功能。GPS 模块集成度高，定位精度高。除了 Android 系统提供了基础的 API 以实现经纬度数等据的获取外，现在很多地图公司(如百度地图与高德地图)都开发了相应的 SDK，提供了很强大的地图服务功能，例如线路规划、地址解析、添加覆盖物、POI 搜索等功能，这为我们开发应用提供很大的帮助。

12.1　支持 GPS 的核心 API

类似于前面讲到的多媒体支持中，Android 提供了一个 AudioManager 服务类。同样，Android 为支持 GPS 提供了一个 LocationManager 服务类。与 GPS 相关的服务，都需要通过这个类来获取。获取该类的方式也与获取 AudioManager 类似，需要调用系统服务。方法如下所示：

```
LocationManager lm = getService(Context.LOCATION_SERVICE);
```

LocationManager 类中提供了丰富的定位服务，其常用方法如表 12.1 所示。

表 12.1　LocationManager 类的常用方法

类型	方法	简述
boolean	addGpsStatusListener(GpsStatus.Listener listener)	添加 GPS 状态监听器
	isProviderEnabled(String provider)	判断指定名称的 LocationProvider 是否可用
void	addProximityAlert(double latitude, double longitude, float radius, long expiration, PendingIntent intent)	添加一个临近警告
	removeGpsStatusListener(GpsStatus.Listener listener)	删除 GPS 状态监听器
	removeProximityAlert(PendingIntent intent)	删除一个临近警告
	requestLocationUpdates(String provider, long minTime, float minDistance, PendingIntent intent)	通过指定的 LocationProvider 周期性地获取定位信息，并通过 intent 启动相应的组件
	requestLocationUpdates(String provider, long minTime, float minDistance, LocationListener listener)	通过指定的 LocationProvider 周期性地获取定位信息，并触发 listener 所对应的触发器
List<String>	getAllProviders()	获取所有的 LocationProvider 列表
	getProviders(Criteria criteria, boolean enabledOnly)	根据指定条件获取满足该条件的全部 LocationProvider 的名称
	getProviders(boolean enabledOnly)	获取所有可用的 LocationProvider
String	getBestProvider(Criteria criteria, boolean enabledOnly)	根据指定条件返回最优的 LocationProvider 对象
GpsStatus	getGpsStatus(GpsStatus status)	获取 GPS 状态
Location	getLastKnownLocation(String provider)	根据 LocationProvider 获取最近一次已知的 Location
LocationProvider	getProvider(String name)	根据名称来获取 LocationProvider

可以从表 12.1 中看到一个重要的类 LocationProvider(定位提供者)。这个类提供了获取定

位组件信息的功能,以判断手机中的定位组件能提供哪些数据,如能否获取海拔信息等。它的常用方法如表 12.2 所示。

表 12.2 LocationProvider 类的常用方法

类型	方法	简述
boolean	hasMonetaryCost()	返回该 LocationProvider 是收费还是免费
	meetsCriteria(Criteria criteria)	判断该 LocationProvider 是否满足 Criteria 条件
	requiresCell()	判断该 LocationProvider 是否需要访问网络基站
	requiresNetwork()	判断该 LocationProvider 是否需要网络数据
	requiresSatellite()	判断该 LocationProvider 是否需要访问基于卫星的定位系统
	supportsAltitude()	判断该 LocationProvider 是否支持高度信息
	supportsBearing()	判断该 LocationProvider 是否支持方向信息
	supportsSpeed()	判断该 LocationProvider 是否支持速度信息
int	getAccuracy()	返回该 LocationProvider 的精度
	getPowerRequirement()	获取该 LocationProvider 的电源需求
String	getName()	返回该 LocationProvider 的名称

最后可以通过 Location 类来获取 GPS 所提供的各项数据,例如精确的经纬度信息、速度信息等。Location 类的常用方法如表 12.3 所示。

表 12.3 Location 类的常用方法

类型	方法	简述
boolean	hasAccuracy()	判断该定位信息是否有精度信息
	hastAltitude()	判断该定位信息是否有高度信息
	hasBearing()	判断该定位信息是否有方向信息
	hasSpeed()	判断该定位信息是否有速度信息
double	getAltitude()	获取定位信息的高度
	getLatitude()	获取定位信息的纬度
	getLongitude()	获取定位信息的经度
float	getAccuracy()	获取定位信息的精度
	getBearing()	获取定位信息的方向
	getSpeed()	获取定位信息的速度
String	getProvider()	获取提供该定位信息的 LocationProvider

在实际应用中,获取 GPS 定位信息的方法主要是以下几个步骤。
- 获取系统的 GPS 服务类,即 LocationManager 实例对象。
- 获取 LocationProvider,定义指定的定位信息,启动定位服务。
- 最后在监听中使用 Location 获取详细的定位信息。

12.2 获取 LocationProvider

通过上一节的介绍,我们知道 GPS 的定位信息是由 LocationProvider 来决定的。这时,如果不做设置,那么 Android 就会提供默认的定位信息服务。但我们也可以通过对 LocationProvider 设置以获取我们需要的服务。下面就介绍几种获取 LocationProvider 的方法。

1)获取所有的 LocationProvider

【例 12.1】 获取所有的 LocationProvider 实例

在 LocationManager 类中提供了 getAllProviders()方法,可以得到所有的 LocationProvider。下面通过代码说明如何获取所有的 LocationProvider。在主界面中通过一个 ListView 显示了获取到的 LocationProvider,布局界面代码十分简单,这里不再展示。

```java
package com.example.locationprovider
import android.content.Context;
import android.location.LocationManager;
import android.support.v7.app.AppCompatActivity;
import android.os.Bundle;
import android.widget.ArrayAdapter;
import android.widget.ListView;
import java.util.List;
public class ProviderList extends AppCompatActivity {
    LocationManager lm;
    ListView lvProviderList;
    @Override
    protected void onCreate(Bundle savedInstanceState) {
        super.onCreate(savedInstanceState);
        setContentView(R.layout.activity_provider_list);
        lm = (LocationManager)getSystemService(Context.LOCATION_SERVICE);
                                            //获取系统服务
        lvProviderList = (ListView)findViewById(R.id.listViewProvider);
        List<String> providerList = lm.getAllProviders();
                                            //获取 Provider,保存在 List 中
        ArrayAdapter<String> adapter = new ArrayAdapter<String>(this,
          android.R.layout.simple_expandable_list_item_1,providerList;
        lvProviderList.setAdapter(adapter);
    }
}
```

在 Genymotion 模拟器中运行该程序,可以看到模拟器提供了 passive、gps 两种 LocationProvider,如图 12.1 所示。在真实手机上运行该程序时可以看到支持的 LocationProvider 又多了一种,即 network。因为手机可以借助移动蜂窝网络或者 WiFi 进行基站定位,这是 Genymotion 模拟器所没有的。

图 12.1 Genymotion 与手机提供的 GPS 服务

下面说明这几种 LocationProvider 的含义。
- **passive**:由 LocationManager.PASSIVE_PROVIDER 常量表示。

- **gps**：表示通过 GPS 获取定位信息的 LocationProvider 对象，由 LocationManager.GPS_PROVIDER 常量表示。
- **network**：表示通过网络获取定位信息的 LocationProvider 对象，由 LocationManager.NETWORK_PROVIDER 常量表示。

在实际应用中，gps 与 network 是最常用的。

2) 通过名称获取 LocationProvider

由上面的例子可以看出，通过 getAllProviders() 方法可以获取所有的 LocationProvider。也可以直接指定 LocationProvider 的名称：使用 LocationManager 所提供的 getProvider(String name) 方法，直接获取需要的 LocationProvider，如下所示：

```
LocationProvider lp = Locationmanager.getProvider
        (LocationManager.GPS_PROVIDER);
```

3) 通过 Criteria 获取 LocationProvider

通过上面两种方法可以直接获取 LocationProvider，但在实际应用中，我们往往对定位服务有一定的要求。例如，必须使 LocationProvider 能够提供海拔信息，这时就需要 Criteria 类，顾名思义就是过滤条件。通过这个类，我们可以选取所需要的 LocationProvider。Criteria 类的常用方法如表 12.4 所示。

表 12.4 Criteria 类的常用方法

方法	简述
setAccuracy(int accuracy)	设置对 LocationProvider 的精度要求
setAltitudeRequired(boolean altitudeRequired)	设置要求 LocationProvider 能提供的高度信息
setBearingRequired(boolean bearingRequired)	设置要求 LocationProvider 能提供的方向信息
setSpeedRequired(boolean speedRequired)	设置要求 LocationProvider 能提供的速度信息
setCostAllowed(boolean costAllowed)	设置 LocationProvider 是否免费
setPowerRequirement(int level)	设置 LocationProvider 的耗电量

在例 12.1 中，可以通过 Criteria 类设置 GPS 能够提供海拔、高精度、方位信息。再通过 getBestProvider 的方法就能获得最佳的 LocationProvider，相关代码如下所示：

```
Criteria criteria = new Criteria();
criteria.setAccuracy(Criteria.ACCURACY_FINE);
criteria.setAltitudeRequired(true);          //要求能提供海拔信息
criteria.setBearingRequired(true);           //要求能提供方位信息
criteria.setCostAllowed(false);              //要求是免费的
criteria.setPowerRequirement(Criteria.POWER_LOW);   //要求是低功耗的
String provider = lm.getBestProvider(criteria, true);  //获取最佳的 Provider
```

12.3 获取定位信息

在了解了 GPS 的相关接口之后，可以编写一个小程序，用于获取定位信息。在本例中，实现了初始化 GPS、获取经纬度、海拔、速度等信息，代码如下所示。

【例 12.2】通过 GPS 获取定位信息实例

```
package com.example.location;
import android.content.Context;
import android.content.Intent;
```

```java
import android.location.Location;
import android.location.LocationListener;
import android.location.LocationManager;
import android.provider.Settings;
import android.support.v7.app.AppCompatActivity;
import android.os.Bundle;
import android.widget.TextView;
import android.widget.Toast;

public class MainActivity extends AppCompatActivity {
    private TextView tvLag,tvAlt,tvSpeed;
    private LocationManager lm = null;
    private Location mLocation;
    private MyLocationListner mLocationListner;
    @Override
    protected void onCreate(Bundle savedInstanceState) {
        super.onCreate(savedInstanceState);
        setContentView(R.layout.activity_main);
        tvLag = (TextView) findViewById(R.id.tvLag);
        tvAlt = (TextView) findViewById(R.id.tvAlt);
        tvSpeed = (TextView) findViewById(R.id.tvSpeed);
        lm = (LocationManager)getSystemService(Context.LOCATION_SERVICE);
        initLocation();
    }
    private void initLocation(){
        //判断GPS是否正常启动
        if(!lm.isProviderEnabled(LocationManager.GPS_PROVIDER)){
            Toast.makeText(MainActivity.this, "请开启GPS...",
                    Toast.LENGTH_SHORT);
            //返回开启GPS导航设置界面
            Intent intent = new Intent(Settings
                    .ACTION_LOCATION_SOURCE_SETTINGS);
            startActivityForResult(intent,0);
            return;
        }
        if (mLocationListner == null) {
            mLocationListner = new MyLocationListner();}
    //注意与Eclipse不同,AndroidSdudio加入了运行时权限机制,需要用try/catch
                                            //抛出异常
        try{
            mLocation = lm.getLastKnownLocation(lm.GPS_PROVIDER);
            updateView(mLocation);
        }catch (SecurityException se){ }
        try{
            //1s更新一次 最小位移变化为2m
            lm.requestLocationUpdates(LocationManager.GPS_PROVIDER,
                    3000, 1, mLocationListner);
        }catch (SecurityException se){
        }
    }
```

```
private class MyLocationListner implements LocationListener
{
    @Override
    public void onLocationChanged(Location location){
        updateView(location); }
    @Override
    public void onProviderDisabled(String provider){
        updateView(null); }
    @Override
    public void onProviderEnabled(String provider) {
        try{
            updateView(lm.getLastKnownLocation(provider));
        }catch (SecurityException e){}}
    @Override
    public void onStatusChanged(String provider, int status, Bundle extras){
        }
}
private void updateView(Location location)
{
    if (location!=null) {
        tvLag.setText("当前经纬度: " + location.getLongitude() + "," +
                    location.getLatitude());
        tvAlt.setText("当前海拔: " + location.getAltitude() + "m");
        tvSpeed.setText("当前速度: " + location.getSpeed() + "m/s");
    }else{
        tvLag.setText("当前经纬度: ");
        tvAlt.setText("当前海拔: " );
        tvSpeed.setText("当前速度: ");
}}}
```

在 Genymotion 模拟器中，免费版提供了 GPS 设置功能，可以方便地设置经度、纬度信息与海拔信息。在 Genymotion 中开启 GPS 如图 12.2 所示，手动输入经纬度，程序运行效果如图 12.3 所示。

图 12.2　在 Genymotion 中开启 GPS

图 12.3　程序运行效果

12.4 使用百度地图服务

在之前的介绍中，我们能够通过 GPS 服务获取到精确的定位信息。但在实际的使用中，我们还需要结合地图，为用户带来更多的定位服务，例如地图定位、线路规划、导航等功能。现在，互联网公司在开发其地图应用的同时，也为开发者开放了 SDK，例如谷歌地图、百度地图与高德地图等。我们在编程时仅需导入库文件，然后调用接口，就可以快捷地将这些地图服务功能集成到自己的应用中。这一节主要介绍如何使用百度地图所提供的 SDK，并实现一个轨迹记录的小应用。

12.4.1 获取百度地图 SDK

百度地图为开发者提供了完善的服务指南，进入百度开放平台：http://lbsyun.baidu.com/，选择"开发"→"Android 开发"→"Android 定位 SDK"，进入页面后，就可以开始准备工作了。

1. 获取密钥

在使用百度地图 SDK 提供的各种功能之前，需要获取百度地图移动版的开发密钥，并且在工程中进行设置，下面介绍设置流程。

登录 API 控制台，单击"创建应用"，如图 12.4 所示。

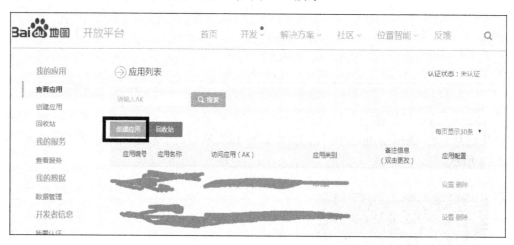

图 12.4 创建应用

在进入页面后选择应用类型，然后填写各类信息，如图 12.5 所示。

在这个页面中需要填写数字签名(SHA1)。以前在 Eclipse 中可以很方便地查看到数字签名，而在 Android Studio 中却不能直接查看。我们可以在 Android Studio 中打开 Terminal，然后定位到密钥所在的文件目录下(通常默认在 C:\Users\Xxx\.android 中)，输入"keytool -v -list -keystore debug.keystore"。其中的"debug.keystore"就是你所使用的调试密钥，然后就可以看到数字签名，如图 12.6 所示。

如果打包生成 APK，则需要单击"Build→Generate Signed"，生成密钥之后再用此方法获得数值签名。

第 12 章　GPS 应用开发与百度地图

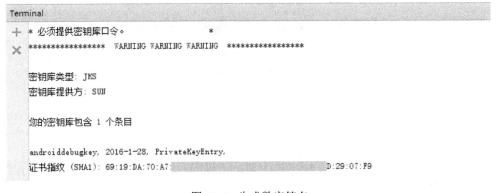

图 12.5　填写信息

图 12.6　生成数字签名

表格中的"包名"可以在 AndroidManifest.xml 文件中查看，如图 12.7 所示。

图 12.7　查看包名

将对应的 SHA1 码与包名填入即可创建应用获得 appkey，如图 12.8 所示。

图 12.8　生成 appkey

2．工程配置

在下载 SDK 库文件之后，就可以开始工程配置了，步骤如下：

首先下载解压开发包，将工程视图改为 ProjectFile，将开发包中的 libs 文件复制到项目名 /app/libs 文件夹中，AndroidStudio 会自动加载 jar 包与 so 文件。然后，右键选择"Add As Library"，导入到工程中。在 src/main 文件目录下创建 jniLibs 文件夹放入对应平台的 so 文件中，工程结构如图 12.9 所示。

图 12.9 工程结构

注意：如果需要在 Genymotion 模拟器上调试该程序，还需要将 x86 平台中的 so 文件导入 jniLibs 中，如图 12.9 所示。

最后在打包混淆的时候，需要注意与地图 SDK 相关的方法不可被混淆。混淆方法如下：

```
-keep class com.baidu.** {*;}
-keep class vi.com.** {*;}
-dontwarn com.baidu.**
```

12.4.2 使用百度地图定位

下面将介绍如何加载百度地图，并调用百度定位 SDK 进行精确定位，与之前准备工作稍有不同的是，这次还需要加入百度定位的 SDK 到工程目录中，才能使用到百度地图定位服务。

【例 12.3】 使用百度地图获取定位信息

首先配置 Manifest 文件，添加 appkey 与所需权限。

```
<?xml version="1.0" encoding="utf-8"?>
<manifest xmlns:android="http://schemas.android.com/apk/res/android"
    package="com.example.bdmap">
    <!--添加需要的权限-->
<uses-permission android:name=
"android.permission.ACCESS_NETWORK_STATE"/>
    <uses-permission android:name="android.permission.INTERNET"/>
<uses-permission android:name=
```

```xml
        "com.android.launcher.permission.READ_SETTINGS" />
    <uses-permission android:name="android.permission.WAKE_LOCK"/>
    <uses-permission android:name="android.permission.CHANGE_WIFI_STATE" />
    <uses-permission android:name="android.permission.ACCESS_WIFI_STATE" />
    <uses-permission android:name="android.permission.GET_TASKS" />
<uses-permission android:name=
"android.permission.WRITE_EXTERNAL_STORAGE"/>
<uses-permission android:name="android.permission.WRITE_SETTINGS" />
<!-- 这个权限用于进行网络定位 -->
<uses-permission android:name=
"android.permission.ACCESS_COARSE_LOCATION" />
<!-- 这个权限用于访问 GPS 定位 -->
<uses-permission android:name=
"android.permission.ACCESS_FINE_LOCATION" />
<!-- 用于访问 wifi 网络信息，wifi 信息会用于进行网络定位 -->
<uses-permission android:name="android.permission.ACCESS_WIFI_STATE" />
    <application
        android:allowBackup="true"
        android:icon="@mipmap/ic_launcher"
        android:label="@string/app_name"
        android:supportsRtl="true"
        android:theme="@style/AppTheme">
        <!--添加 appkey-->
        <meta-data
            android:name="com.baidu.lbsapi.API_KEY"
            android:value="7znnXUp2G3LIMQ8g3z7OSKZp" />
        <activity android:name=".BaiduActivity">
            <intent-filter>
                <action android:name="android.intent.action.MAIN" />
                <category android:name="android.intent.category.LAUNCHER" />
            </intent-filter>
        </activity>
        <!--添加百度服务-->
        <service
            android:name="com.baidu.location.f"
            android:enabled="true"
            android:process=":remote" >
        </service>
    </application>
</manifest>
```

然后定义 Baidu Activity：

```java
package com.example.bdmap;
import android.support.v7.app.AppCompatActivity;
import android.os.Bundle;
import com.baidu.location.BDLocation;
import com.baidu.location.BDLocationListener;
import com.baidu.location.LocationClient;
```

```java
import com.baidu.location.LocationClientOption;
import com.baidu.mapapi.SDKInitializer;
import com.baidu.mapapi.map.BaiduMap;
import com.baidu.mapapi.map.MapStatus;
import com.baidu.mapapi.map.MapStatusUpdateFactory;
import com.baidu.mapapi.map.MapView;
import com.baidu.mapapi.map.MyLocationData;
import com.baidu.mapapi.model.LatLng;
public class BaiduActivity extends AppCompatActivity {
    MapView mMapView = null;
    BaiduMap mBaiduMap;
    LocationClient mLocClient;
    boolean isFirstLoc = true;        //是否首次定位
    public MyLocationListenner myListener = new MyLocationListenner();
    @Override
    protected void onCreate(Bundle savedInstanceState) {
        super.onCreate(savedInstanceState);
        SDKInitializer.initialize(getApplicationContext());
        setContentView(R.layout.activity_baidu);
        mMapView = (MapView) findViewById(R.id.bmapView);
        mBaiduMap = mMapView.getMap();
        //开启定位图层
        mBaiduMap.setMyLocationEnabled(true);
        //定位初始化
        mLocClient = new LocationClient(this);
        mLocClient.registerLocationListener(myListener);
        LocationClientOption option = new LocationClientOption();
        option.setOpenGps(true);              //打开gps
        option.setCoorType("bd09ll");         //设置坐标类型
        option.setScanSpan(1000);
        mLocClient.setLocOption(option);
        mLocClient.start();
    }
    //定位SDK监听函数
    public class MyLocationListenner implements BDLocationListener {

        @Override
        public void onReceiveLocation(BDLocation location) {
            //map view 销毁后不在处理新接收的位置
            if (location == null || mMapView == null) {
                return;
            }
            MyLocationData locData = new MyLocationData.Builder()
                    .accuracy(location.getRadius())
                    //此处设置开发者获取到的方向信息,顺时针0~360
                    .direction(100).latitude(location.getLatitude())
                    .longitude(location.getLongitude()).build();
```

```
            mBaiduMap.setMyLocationData(locData);
            if (isFirstLoc) {
                isFirstLoc = false;
                LatLng ll = new LatLng(location.getLatitude(),
                        location.getLongitude());
                MapStatus.Builder builder = new MapStatus.Builder();
                builder.target(ll).zoom(18.0f);
            mBaiduMap.animateMapStatus(
            MapStatusUpdateFactory.newMapStatus(builder.build()));
             }
        }
        public void onReceivePoi(BDLocation poiLocation) {
        }
    }
    @Override
    protected void onDestroy() {
        super.onDestroy();       //在 activity 执行 onDestroy 时执行
                                 mMapView.onDestroy(),实现地图生命周期管理
        mLocClient.stop();
        mBaiduMap.setMyLocationEnabled(false);   //关闭定位图层
        mMapView.onDestroy();
        mMapView = null;
        super.onDestroy();
    }
    @Override
    protected void onResume() {
        super.onResume();        //在 activity 执行 onResume 时执行
                                 mMapView.onResume(),实现地图生命周期管理
        mMapView.onResume();
    }
    @Override
    protected void onPause() {
        super.onPause();         //在 activity 执行 onPause 时执行
                                 mMapView.onPause(),实现地图生命周期管理
        mMapView.onPause();
    }
}
```

最后在布局文件中添加地图控件即可。

```
<com.baidu.mapapi.map.MapView
    android:id="@+id/bmapView"
    android:layout_width="fill_parent"
    android:layout_height="fill_parent"
    android:clickable="true" />
```

在 Genymotion 模拟器中运行程序,会自动定位到网络所在地址,效果如图 12.10 所示:

图 12.10　定位效果

第 2 篇
综合项目篇

第 13 章　基于 Android 的计算器
第 14 章　图形化数字游戏
第 15 章　基于 Android 的音频播放器
第 16 章　Android 数据库记事本

第 13 章　基于 Android 的计算器

13.1　系统分析

计算器是一种在日常生活中应用广泛的电子产品，无论是在超市商店，还是在办公室，或是家庭都有着它的身影。在智能手机普及的今天，基于 Android 电子计算器的运用也非常普遍。通过该软件，用户能够随时随地地进行通用计算，极大地方便使用者。

可行性分析
(1) 要求
软件的功能要符合使用者的基本情况。可以方便地进行加、减、乘、除等运算操作。系统的界面要简易、快捷，不要有过多的修饰或者不必要的功能。软件要有一定的安全性和健壮性，不能影响手机系统的运行。
(2) 目标
方便所有对计算器功能有需求的手机用户。
(3) 评价尺度
软件要在 5 个工作日内开发完毕并交付使用，"用户"需要在 1 个工作日内确认需求文档，去除其中的不合格因素，程序员需要在"用户"确认需求后 3 个工作日内完成系统设计、详细设计、编码、测试。最后 1 个工作日在"用户"的协助下，完成安装和反馈修改等工作。

13.2　系统设计

13.2.1　系统目标

根据用户对 Android 计算器的使用要求，制定系统目标如下。
- 操作简单、易于掌握，界面简洁清爽。
- 方便地进行加、减、乘、除等操作。
- 要包含小数点运算和输入回退功能。
- 能够进行多次叠加运算。
- 系统运行稳定，不能和手机固有软件冲突，安全可靠。

13.2.2　系统功能结构

Android 计算器的功能结构图如图 13.1 所示。

第 13 章 基于 Android 的计算器

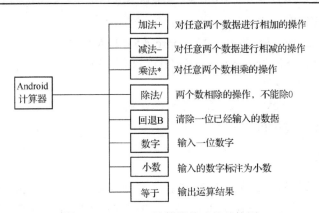

图 13.1 Android 计算器的功能结构图

13.2.3 系统业务流程

Android 计算器软件的逻辑流程图如图 13.2 所示。

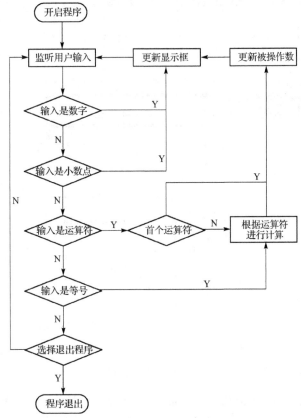

图 13.2 Android 计算器软件的逻辑流程图

13.3 系统实施

13.3.1 开发及运行环境

本项目的软件开发及运行环境如下。

- 操作系统：Windows 7。
- 开发工具：Android Studio 1.1.0 + Android 5.0.1（API 21）。
- JDK 环境：Java SE Development Kit（JDK）1.8.0_31。
- 开发语言：Java、XML。
- 运行平台：AVD Nexus S API 21（虚拟机设置）。
- 分辨率：480×800 hdpi。

13.3.2 项目的创建

Android 计算器系统的项目名称为 MyCalc，该项目是使用 Android Studio 1.1.0 + Android 5.0.1（API 21）开发的，在 Android Studio 开发环境中创建该项目的步骤如下。

（1）启动 Android Studio，经过初始化配置之后，出现如图 13.3 所示的界面。在这个界面可以新建项目，也可以导入本地的项目等，在左侧可以查看最近打开的项目。这里直接选择新建项目。

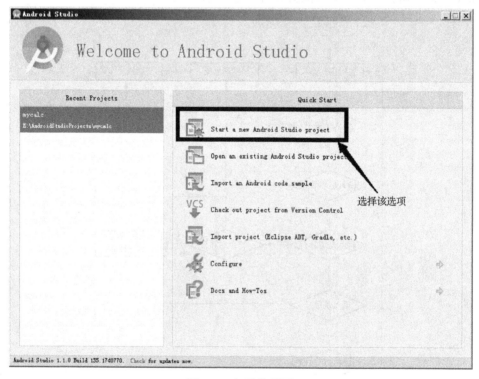

图 13.3 初始化界面

（2）单击"Start a new Android Studio project（新建项目）"选项后，会弹出 Create New Project 窗口。在该窗口中首先输入项目名 MyCalc，并且输入公司域名为 test.com，然后输入项目存储的物理路径（如图 13.4 所示），最后单击"Next"按钮。

（3）弹出的"Target Android Devices"页面支持用户适配 TV、Wear、Glass 等，这里只选择第一项"Phone and Tablet"，然后选好项目支持的最小 SDK，最后单击"Next"按钮，如图 13.5 所示。

（4）弹出的"Add Activity"页面中选择一个 Activity 模板添加到工程中，和 Eclipse 很像，

第 13 章 基于 Android 的计算器

直接选择 "Blank Activity" 创建一个 "Hello World!" 程序，然后单击 "Next" 按钮，如图 13.6 所示。

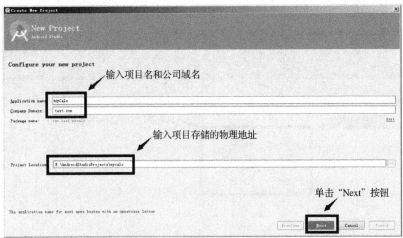

图 13.4 输入项目名称和路径等

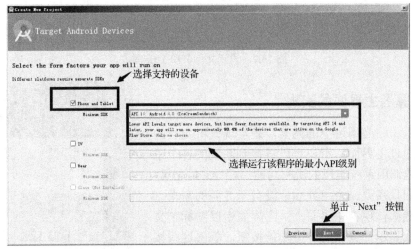

图 13.5 选择设备和 Android 版本

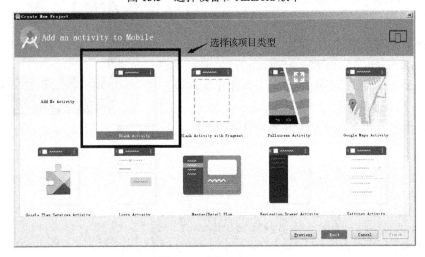

图 13.6 添加 Activity

(5)在弹出的"Customize the Activity"页面中可以设置主 Activity 的名称、XML 界面配置文件名称、项目的 Title 和菜单名称,然后单击"Finish"按钮完成工程的创建,如图 13.7 所示。

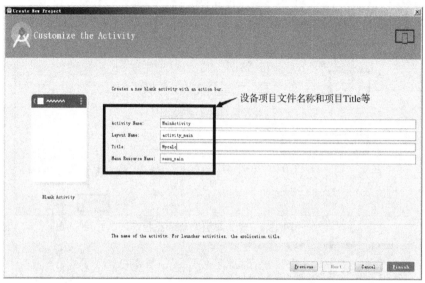

图 13.7　设置文件名和 Title 等

13.3.3　计算器主界面的实现

主界面是程序必不可少的元素,是实现人机交互的必有环节。通过主界面,用户可以链接到系统的各个子模块,快速地调用系统实现的所有功能。在 Android 计算器中,我们只需要一个主界面,没有子界面的需求,所以所有功能都体现在主界面上。主界面中以图标和文字结合的方式显示计算器的各个功能按钮,单击这些按钮,完成计算器的输入和计算功能。Android 计算器主界面如图 13.8 所示。

1. 设计界面布局 XML 文件

在 Android Studio 的工程中 res\layout\目录下,找到 activity_main.xml 文件,这是用来作为主界面布局的设置文件。在该布局文件中,添加一个 AbsoluteLayout(绝对布局管理器)组件,用于显示功能图标和图标上的文本,其代码如下:

图 13.8　Android 计算器主界面

```xml
<?xml version="1.0" encoding="utf-8"?>
<AbsoluteLayout android:id="@+id/widget0"
  android:layout_width="fill_parent"
  android:layout_height="fill_parent"
   xmlns:android="http://schemas.android.com/apk/res/android">
</AbsoluteLayout>
```

2. 添加控件并绑定响应函数

在 AbsoluteLayout(绝对布局管理器)中，定义一个 TextView 用于显示中间数字和结果，定义 17 个 Button 用于输入 0~9 的数字和+、-、*、/、=、.运算符号以及"Back"输入回退按钮，并用"android:onClick"属性绑定它们的响应函数，具体代码如下。

```
<TextView android:id="@+id/txtResult" android:layout_width="460px"
    android:layout_height="80px" android:background="#ffffffff"
    android:text="0.0" android:textSize="28sp" android:textStyle="bold"
    android:textColor="#ff333333" android:layout_x="11dp"
    android:layout_y="16dp" android:gravity="right" android:padding="2px" />
<Button android:id="@+id/btn1" android:layout_width="80px"
    android:layout_height="80px" android:text="1" android:textSize="20sp"
    android:background="#ff5599ff" android:textStyle="bold"
    android:gravity="center" android:layout_x="15dp" android:layout_y="85dp"
    android:onClick="digital_click" />

<Button android:id="@+id/btn2" android:layout_width="80px"
    android:layout_height="80px" android:text="2" android:textSize="20sp"
    android:background="#ff5599ff" android:textStyle="bold"
    android:gravity="center" android:layout_x="85dp" android:layout_y="85dp"
    android:onClick="digital_click" />

<Button android:id="@+id/btn3" android:layout_width="80px"
    android:layout_height="80px" android:text="3" android:textSize="20sp"
    android:background="#ff5599ff" android:textStyle="bold"
    android:gravity="center" android:layout_x="155dp" android:layout_y="85dp"
    android:onClick="digital_click" />

<Button android:id="@+id/btnAdd" android:layout_width="80px"
    android:layout_height="80px" android:text="+" android:textSize="20sp"
    android:textStyle="bold" android:gravity="center" android:layout_x="225dp"
    android:layout_y="85dp" android:onClick="add" />

<Button android:id="@+id/btn4" android:layout_width="80px"
    android:layout_height="80px" android:text="4" android:textSize="20sp"
    android:background="#ff5599ff" android:textStyle="bold"
    android:gravity="center" android:layout_x="15dp" android:layout_y="155dp"
    android:onClick="digital_click" />

<Button android:id="@+id/btn5" android:layout_width="80px"
    android:layout_height="80px" android:text="5" android:textSize="20sp"
    android:background="#ff5599ff" android:textStyle="bold"
    android:gravity="center" android:layout_x="85dp" android:layout_y="155dp"
    android:onClick="digital_click" />

<Button android:id="@+id/btn6" android:layout_width="80px"
    android:layout_height="80px" android:text="6" android:textSize="20sp"
    android:background="#ff5599ff" android:textStyle="bold"
```

```xml
        android:gravity="center" android:layout_x="155dp" android:layout_y="155dp"
        android:onClick="digital_click" />

<Button android:id="@+id/btnsub" android:layout_width="80px"
android:layout_height="80px" android:text="-" android:textSize="20sp"
        android:textStyle="bold" android:gravity="center" android:layout_x="225dp"
        android:layout_y="155dp" android:onClick="sub" />

<Button android:id="@+id/btn7" android:layout_width="80px"
android:layout_height="80px" android:text="7" android:textSize="20sp"
        android:background="#ff5599ff" android:textStyle="bold"
        android:gravity="center" android:layout_x="15dp" android:layout_y="225dp"
        android:onClick="digital_click" />

<Button android:id="@+id/btn8" android:layout_width="80px"
android:layout_height="80px" android:text="8" android:textSize="20sp"
        android:background="#ff5599ff" android:textStyle="bold"
        android:gravity="center" android:layout_x="85dp" android:layout_y="225dp"
        android:onClick="digital_click" />

<Button android:id="@+id/btn9" android:layout_width="80px"
android:layout_height="80px" android:text="9" android:textSize="20sp"
        android:background="#ff5599ff" android:textStyle="bold"
        android:gravity="center" android:layout_x="155dp" android:layout_y="225dp"
        android:onClick="digital_click" />

<Button android:id="@+id/btnMul" android:layout_width="80px"
android:layout_height="80px" android:text="*" android:textSize="20sp"
        android:textStyle="bold" android:gravity="center" android:layout_x="225dp"
        android:layout_y="225dp" android:onClick="mul" />

<Button android:id="@+id/btn0" android:layout_width="130px"
android:layout_height="80px" android:text="0" android:textSize="20sp"
        android:textStyle="bold" android:gravity="center" android:layout_x="15dp"
        android:layout_y="295dp" android:onClick="digital_click" />

<Button android:id="@+id/btnPoint" android:layout_width="80px"
android:layout_height="80px" android:text="." android:textSize="20sp"
        android:textStyle="bold" android:gravity="center" android:layout_x="155dp"
        android:layout_y="295dp" android:onClick="point_click" />

<Button android:id="@+id/btnDiv" android:layout_width="80px"
android:layout_height="80px" android:text="/" android:textSize="20sp"
        android:textStyle="bold" android:gravity="center" android:layout_x="225dp"
        android:layout_y="295dp" android:onClick="div" />
<Button android:id="@+id/btndel" android:layout_width="80px"
        android:layout_height="85px" android:text="back" android:textSize="15sp"
```

```
        android:textStyle="bold" android:textColor="#ffff0000"
        android:gravity="center" android:layout_x="15dp" android:layout_y="365dp"
        android:onClick="del" />

<Button android:id="@+id/btnequ" android:layout_width="290px"
    android:layout_height="80px" android:text="=" android:textSize="20sp"
        android:textStyle="bold" android:gravity="center" android:layout_x="85dp"
        android:layout_y="365dp" android:onClick="equ" />
```

13.3.4 计算器逻辑的实现

由于计算器本身的功能较为单一，实现简单，所以用一个 java 文件即可实现对所有功能的编制。建立工程后，将默认的主程序文件，也就是 Android Studio 工程界面中的 java 文件夹内的"com.test.mycalc"包中的"MainActivity.java"文件进行编辑。该文件所在位置如图 13.9 所示，下面将该文件分模块解析。

1．文件导入的包和类

由于程序需要使用到库函数中的模块定义，所以在编写程序时，需"import（导入）"相应的内容来支持程序中对该类的使用。例如要使用 Android Studio 中的"按钮元件，就必须导入"Button"类，Android 计算器的文件头如下：

图 13.9 Android 源程序文件

```java
package com.test.mycalc ;                   //声明程序所属包
import java.text.NumberFormat;              //导入 NumberFormat 类
import android.app.Activity;                //导入 Activity 类
import android.os.Bundle;                   //导入 Bundle 类
import android.view.View;                   //导入 View 类
import android.widget.Button;               //导入 Button 类
import android.widget.TextView;             //导入 TextView 类
```

2．类的定义

构造函数是程序初次运行时自动调用的函数，一般在这里对程序中的变量、界面元素等做生成和初始化，Android 计算器的初始化单元如下：

```java
public class MainActivity extends Activity {
    /** Called when the activity is first created. */

    double firstNum = 0;                    //第一个输入的数据
    char currentSign = '+';                 //记录第一次输入的符号
    StringBuffer currentNum = new StringBuffer();//得到 textview 中的数据
    boolean isFirstPoint = false;           //第一个数据是否是小数点
    TextView txtResult;                     //输出结果


    @Override
    public void onCreate(Bundle savedInstanceState) {
```

```
        super.onCreate(savedInstanceState);
        setContentView(R.layout.activity_main);

        txtResult = (TextView) findViewById(R.id.txtResult);
    }
```

3. 子函数设计

程序需求的各种功能，分别编制子函数实现，可以安排团队中合理分工，Android 计算器的子函数设计包括以下内容。

1) 输入框恢复与初始化函数

用于实现数字数组当前值设置为 0，并清除小数点因素。用于被其他子函数调用。详细代码如下：

```
    /**
     * 对数据进行初始化
     *
     * @return
     */
    public void init() {
        currentNum.delete(0, currentNum.length());//设置当前 textView 中的值为 0
        isFirstPoint = false;
    }
```

2) 数值转换函数

用于从输入框的 string 变量中，获取输入的双精度浮点数。计算函数可以调用。详细代码如下：

```
    /**
     * 将输入的数据转换成 double 类型
     *
     * @return
     */
    public double stringToDouble() {
        if (currentNum.length() == 0) {//如果没有输入的数据
            return 0;
        }
        double result = Double.parseDouble(currentNum.toString());
        return result;
    }
```

3) 计算函数

用于实现计算本身，判定进行加、减、乘、除的哪一种运算并执行。运算结果保留 2 位小数。详细代码如下：

```
    /**
     * 进行计算处理
     * @return
     */
    public double calcu(){
```

```
        double result=0;
        switch(currentSign){
            case '+':
                result=firstNum+stringToDouble();
                break;
            case '-':
                result=firstNum-stringToDouble();
                break;
            case '*':
                result=firstNum*stringToDouble();
                break;
            case '/':
                result=firstNum/stringToDouble();
                break;
        }
        //对小数点后的数据进行格式化
        NumberFormat format =NumberFormat.getInstance();
        format.setMaximumFractionDigits(2);
        result=Double.parseDouble(format.format(result));
        return result;
    }
```

4) 显示数据函数

用于刷新文本框的显示内容。供其他函数调用，详细代码如下：

```
    /**
     * 显示数据
     */
    public void display(){
        txtResult.setText(currentNum.toString());
    }
```

5) 数字按钮单击响应函数

用于响应 0~9 数字按钮的单击动作。自动判断单击的按钮并将该数字添加到文本框作为输入的一部分，详细代码如下：

```
    /**
     * 处理数字按钮的单击
     *
     * @param view
     */
    public void digital_click(View view) {
        Button btnDigital=(Button) view;
        char text=btnDigital.getText().charAt(0);
        currentNum.append(text);
        display();
    }
```

6）加法按钮单击响应函数

用于响应加法按钮的单击动作。进行加法运算并更新文本框、结果变量、运算符变量且刷新当前输入数组，详细代码如下：

```java
/**
 * 处理加法
 *
 */
public void add(View view) {
    double result=calcu();
    txtResult.setText(String.valueOf(result));
    firstNum=result;
    currentSign='+';
    init();
}
```

7）减法按钮单击响应函数

用于响应减法按钮的单击动作。进行减法运算并更新文本框、结果变量、运算符变量且刷新当前输入数组，详细代码如下：

```java
/**
 * 处理减法
 *
 */
public void sub(View view) {
    double result=calcu();
    txtResult.setText(String.valueOf(result));
    firstNum=result;
    currentSign='-';
    init();
}
```

8）乘法按钮单击响应函数

用于响应乘法按钮的单击动作。进行乘法运算并更新文本框、结果变量、运算符变量且刷新当前输入数组，详细代码如下：

```java
/**
 * 处理乘法
 *
 */
public void mul(View view) {
    double result=calcu();
    txtResult.setText(String.valueOf(result));
    firstNum=result;
    currentSign='*';
    init();
}
```

9）除法按钮单击响应函数

用于响应除法按钮的单击动作。进行除法运算并更新文本框、结果变量、运算符变量且刷新当前输入数组，详细代码如下：

```java
/**
 * 处理除法
 *
 */
public void div(View view) {
    double result=calcu();
    txtResult.setText(String.valueOf(result));
    firstNum=result;
    currentSign='/';
    init();
}
```

10) 处理等于函数

用于响应不断单击等号按钮的动作。不断单击等号按钮等于累加运算。

```java
/**
 * 处理等于
 *
 */
public void equ(View view) {
    double result=calcu();
    txtResult.setText(String.valueOf(result));
    firstNum=result;
    currentSign='+';
    init();
}
```

11) 小数点按钮单击响应函数

用于响应小数点按钮的单击动作。不进行运算，判断是否为第一个小数点，若不是则将小数点加入到当前输入数字数组中，详细代码如下：

```java
/**
 * 处理小数点
 *
 */
public void point_click(View view) {
    if(isFirstPoint){                      //当第一个数据为小数点时，程序返回
        return;
    }
    if(currentNum.length()==0){    //当没有输入的数据时返回
        return;
    }
    Button btnPoint=(Button)view;
    char text=btnPoint.getText().charAt(0);
    currentNum.append(text);
    isFirstPoint=true;
    display();
}
```

12) 删除按钮单击响应函数

用于响应删除按钮的单击动作。不进行运算，判断是否有数字可删除，若有则删除一位当前输入数据，详细代码如下：

```
/**
 * 处理删除数据
 */
public void del(View view){

    if(currentNum.length()>=1)
    {
        currentNum.delete(currentNum.length()-1,currentNum.length());
    }
    if(currentNum.length()==0){

        init();
        display();
    }
    txtResult.setText(currentNum);
}
```

13.4　系统运行与测试

界面设计和代码设计完成后，单击 Android Studio 开发环境的工具栏中的 ▶ 图标，或者选择菜单栏中的"Run"→"Run"命令，运行该项目，可以调试查看 Android 计算器程序。当用户单击运行后，会弹出设备选择窗口，如图 13.10 所示。在该窗口中，可以选择连接在计算机上的手机设备、运行中的 Android 虚拟机或者新启动一个 Android 虚拟设备。这里选择启动早先使用 AVD 创建的"Nexus S"设备。

在运行好的窗口中输入测试数据"26*3.4="结果"88.4"如图 13.11 所示，项目开发完成！

图 13.10　设备选择窗口

图 13.11　项目运行结果

第14章 图形化数字游戏

14.1 系统分析

手机中的小游戏是一种在日常生活中应用广泛的休闲软件，无论是在超市商店，还是在办公室，或是家庭都有着它的身影。随着移动互联网和智能手机的不断发展和进步，当今市场上已经出现了多种简单轻松的小游戏，几乎每一位智能手机的使用者都会在种类繁多的 App 网站上下载一些小游戏，这些数字小游戏可以使人们在紧张的工作节奏中很好地放松身心。

(1) 要求

小游戏的功能要符合使用者的基本情况，可以方便地进行操作，游戏的界面要求美观，有自己的特色。系统的界面要简易、快捷，不要有过多的修饰或者不必要的功能。游戏流畅，但不能影响手机系统的运行。

(2) 目标

方便所有对休闲娱乐功能有需求的手机用户。

14.2 系统设计

14.2.1 系统目标

根据用户对 Android 数字小游戏的使用要求，制定如下系统目标。
- 操作简单、易于掌握，界面简洁清爽。
- 方便进行游戏的触摸操作。
- 要包含记分和游戏终止功能。
- 系统运行稳定，不能和手机固有软件冲突，安全可靠。

14.2.2 系统功能结构

Android 数字小游戏的功能结构图如图 14.1 所示。

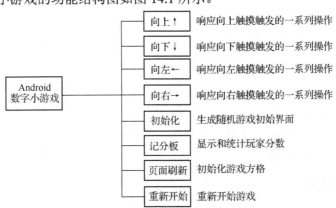

图 14.1 Android 数字小游戏功能结构图

14.2.3 系统业务流程

Android 数字小游戏软件的业务流程图如图 14.2 所示。

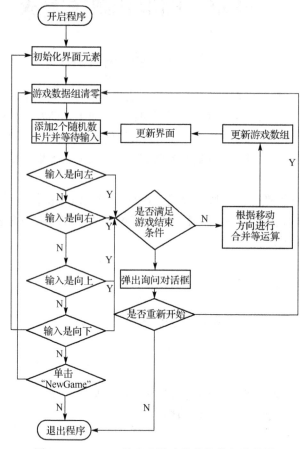

图 14.2 Android 数字小游戏软件的业务流程图

14.3 系统实施

14.3.1 开发及运行环境

本项目的软件开发及运行环境如下。
- 操作系统：Windows 7。
- 开发工具：Android Studio 1.1.0 + Android 4.4.2（API 19）。
- JDK 环境：Java SE Development Kit（JDK）1.8.0_31。
- 开发语言：Java、XML。
- 运行平台：AVD Nexus S API 19（虚拟机设置）。
- 分辨率：480×800 hdpi。

14.3.2 项目的创建

Android 数字小游戏系统的项目名称为 MiniGame，该项目是使用 Android Studio 1.1.0 +

Android 4.4.2(API 19)开发的,在 Android Studio 开发环境中创建该项目的步骤如下。

(1)启动 Android Studio,经过初始化配置之后,出现如图 14.3 所示的界面。在这个界面中可以新建项目,也可以导入本地项目或者 GitHub 上的项目等,在左侧可以查看最近打开的项目等,这里选择新建项目。

图 14.3　初始化界面

(2)单击"Start a new Android Studio project"选项后,会弹出 Create New Project 窗口。在该窗口中首先输入项目名称 MiniGame,并且输入公司域名为 as.com,然后输入项目的存储路径(如图 14.4 所示),最后单击"Next"按钮。

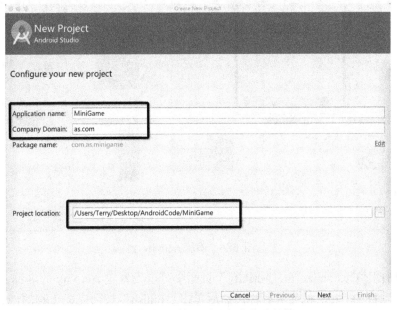

图 14.4　输入项目名称和路径等

(3)弹出的"Target Android Devices"页面支持用户适配 TV、Wear、Glass 等，这里只选择第一项"Phone and Tablet"，然后选好项目支持的最小 SDK（API 19），最后单击"Next"按钮，如图 14.5 所示。

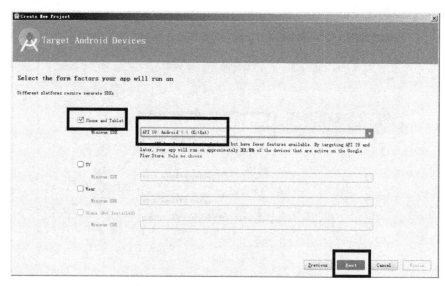

图 14.5 选择设备和 Android 版本

(4)在弹出的"Add Activity"页面中选择一个 Activity 模板添加到工程中，和 Eclipse 很像，这里直接选择"Blank Activity"创建一个"Hello World！"程序，然后单击"Next"按钮，如图 14.6 所示。

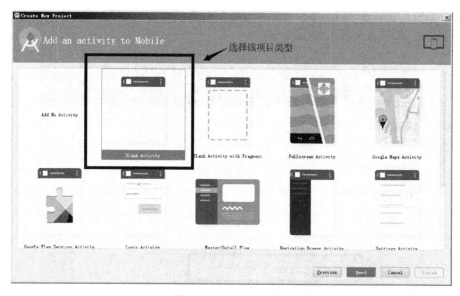

图 14.6 添加 Activity

(5)在弹出的"Customize the Activity"页面中可以设置主 Activity 的名称、XML 界面配置文件名称、项目的 Title 和菜单名称，然后单击"Finish"按钮完成工程的创建，如图 14.7 所示。

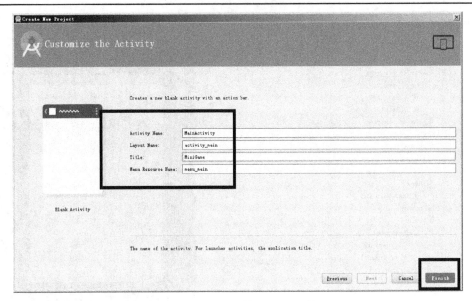

图 14.7　设置文件名和 Title 等

14.3.3　项目工程结构

在我们编写代码之前，首先要制定好项目的系统文件夹的组织结构目录，分门别类地管理我们的项目资源，包括不同的界面、类、数据模型、图片资源等，这样不但可以保证系统开发过程的规范性和程序员的可替代性，还有利于保证团队开发的一致性。创建好系统的文件夹后，在开发的过程中，只需要将我们新创建的类文件或者资源文件、脚本文件等保存到相应的文件夹中即可。Android 数字小游戏项目所使用的文件夹组织结构如图 14.8 所示。

14.3.4　游戏主界面的实现

主界面是程序必不可少的元素，是实现人机交互的必有环节。通过主界面，用户可以链接到系统的各个子模块，快速地调用系统实现的所有功能。在 Android 数字小游戏中，我们只需要一个主界面，没有子界面的需求，所以所有功能都体现在主界面上。主界面中以图标和文字结合的方式显示各个计数小方块，通过上下左右四个方向拖动小方块。完成相同数字的相加和记分，从而进行游戏。Android 数字小游戏主界面如图 14.9 所示。

1．设计界面布局 XML 文件

在 Android Studio 的工程中 res\layout\目录下，找到 activity_main.xml 文件，这是用来作为主界面布局的设置文件，在该布局文件中，添加一个 LinearLayout 组件，用于显示功能图标和图标上的文本，其代码如下：

```xml
<LinearLayout xmlns:android="http://schemas.android.com/apk/res/android"
    xmlns:tools="http://schemas.android.com/tools"
    android:layout_width="match_parent"
    android:layout_height="match_parent"
    android:orientation="vertical"
    tools:context=".MainActivity" >                          //设置关联文件
</LinearLayout>
```

图 14.8　文件夹组织结构　　　　　　　图 14.9　Android 数字小游戏主界面

2．子界面组件属性设置

在主界面中，定义两个子界面(布局)。第一个子界面定义三个 TextView 用于显示顶层的提示、动态分数和"New Game"按钮，具体代码如下：

```
<LinearLayout                                          //定义子界面
    android:layout_width="fill_parent"                 //设置控件 ID
    android:layout_height="wrap_content"
    android:orientation="horizontal"
    >
    <TextView                                          //提示文本框
        android:id="@+id/tv_score_text"                //设置控件 ID
        android:layout_width="wrap_content"
        android:layout_height="wrap_content"
        android:textSize="18dp"
        android:textColor="#000000"
        android:padding="8dp"
        android:text="You Score Is:"                   //设置初始文本
        />
    <TextView                                          //动态显示分数文本框
        android:id="@+id/tv_score_show"                //设置控件属性
        android:layout_width="wrap_content"
        android:layout_height="wrap_content"
        android:textSize="20dp"
        android:textColor="#000000"
        android:padding="10dp"
        android:text="0"                               //设置初始文本
        />
```

```
        <TextView                              //作为"New Game"按钮的文本框
            android:id="@+id/tv_newgame"        //设置控件ID
            android:layout_width="wrap_content"
            android:layout_height="wrap_content"
            android:textSize="18dp"
            android:textColor="#000000"
            android:paddingTop="8dp"
            android:paddingLeft="20dp"
            android:text="NEW Game"             //设置初始文本
            />
    </LinearLayout>
```

第二个子界面为游戏子界面，设置其子界面大小，具体代码如下：

```
    <com.as.minigame.GameView                  //关联到GameView类
        android:id="@+id/gv_show"              //设置控件ID
        android:layout_width="fill_parent"
        android:layout_height="fill_parent"
        android:layout_weight="1"
        >
    </com.as.minigame.GameView>
```

14.3.5 数字小游戏逻辑的实现

由于小游戏涉及图像的更新和分数统计，又要兼顾总体运行。为了方便调试和修改，也为了更好地分工，采用三个java类文件实现对所有功能的编制。建立工程后，除默认的主程序文件，即Android Studio工程界面中的java文件夹内的"com.as.minigame"包中的"MainActivity.java"文件外，再创建两个"Java Class"，分别为"Card.java"和"GameView.java"。文件所在位置如图14.10所示，下面将三个文件分模块解析。

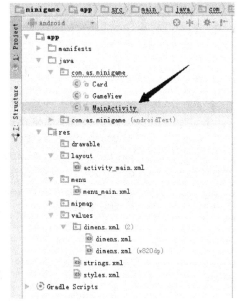

图14.10 Android源程序文件

1. 实现主程序MainActivity

Android数字小游戏工程中的"MainActivity.java"文件主要包含了程序主界面类的定义，以及各个元件的定义和指定界面元素与响应函数的关联。下面详细解析该类的各个模块。

1）文件导入的包和类

由于该类需要使用到库函数中的模块定义，所以在编写程序时，需"import（导入）"相应的内容来支持程序中对该类的使用。例如要使用Android Studio中的"文本"元件，就必须导入"TextView"类，Android数字小游戏的文件头如下：

```
        package com.as.minigame;                //设置包名

        import java.util.Timer;
        import java.util.TimerTask;
```

```java
import android.os.Bundle;
import android.os.Handler;
import android.os.Message;
import android.app.Activity;
import android.view.View;
import android.view.View.OnClickListener;
import android.widget.TextView;
```

2) 类的定义

Android 数字小游戏的初始化单元如下:

```java
public class MainActivity extends Activity {
    TextView score_show;
    GameView gv;
    TextView new_game;
    Handler handler = new Handler(){

        @Override
        public void handleMessage(Message msg) {
            //TODO Auto-generated method stub
            super.handleMessage(msg);
            int num =msg.arg1;
            score_show.setText(num+"");

        }



    };

    protected void onCreate(Bundle savedInstanceState) {
        super.onCreate(savedInstanceState);
        setContentView(R.layout.activity_main);
        score_show = (TextView) findViewById(R.id.tv_score_show);
        gv =(GameView) findViewById(R.id.gv_show);
        new_game =(TextView) findViewById(R.id.tv_newgame);
        new_game.setOnClickListener(new OnClickListener() {

            @Override
            public void onClick(View arg0) {
                gv.GameStart();
                gv.score = 0;


            }
        });
```

3) 程序初始化工作

程序需求的各种初始化,可在主类中编制子函数实现,Android 数字小游戏的子函数设计包括以下内容:

设置计时器 timer 并使用 timer 定时传递包含分数信息的 message，以便界面定时刷新分数。详细代码如下：

```java
Timer timer =new Timer();
timer.schedule(new TimerTask() {

    @Override
    public void run() {
        Message msg = new Message();
        msg.arg1 = gv.score ;
        handler.sendMessage(msg);
    }
}, 80, 150);
score_show.setText(100+"");
```

2. 卡片文件 Card.java 的实现

Android 数字小游戏工程中的"Card.java"文件主要包含了游戏界面的定义，划分出游戏需要的方格，以及各个方块的位置、大小、颜色、初始文字等。下面详细解析该类的各个模块。

1) 文件导入的包和类

由于该类需要使用库函数中的模块定义，所以在编写程序时，需"import"相应的内容来支持程序中对该类的使用。Card.java 文件头如下：

```java
package com.as.minigame;                        //文件所属包

import android.content.Context;
import android.view.Gravity;
import android.widget.FrameLayout;
import android.widget.TextView;
```

2) 实现 Card.java 文件

Card.java 的初始化单元如下：

```java
public class Card  extends FrameLayout {

    private TextView text;
    private int number=0;
    public int getNumber() {
        return number;
    }


    public void setNumber(int number) {
        this.number = number;
        if(number<2){
            text.setText("");
        }else{
```

```java
            if(number>=64){
                text.setTextColor(0xffffff00);
            }else{
                text.setTextColor(0xff000000);
            }
            text.setText(number+"");
        }
    }

    public Card(Context context) {
        super(context);
        //TODO Auto-generated constructor stub
        text = new TextView(context);
        text.setTextSize(28);
        text.setBackgroundColor(0x9966cccc);
        text.setGravity(Gravity.CENTER);
        LayoutParams params = new LayoutParams(-1,-1);
        params.setMargins(10, 10, 0, 0);
        addView(text, params);

    }
}
```

3. 实现游戏交互文件 GameView.java

Android 数字小游戏工程中的"GameView.java"文件主要包含了游戏交互的响应、分数计算、游戏结束的判定等游戏的主要功能的实现，以及游戏界面的初始化和游戏每一步结果的实现和反馈。下面详细解析该类的各个模块。

1) 文件导入的包和类

由于该类需要使用库函数中的模块定义，所以在编写程序时，需"import"相应的内容来支持程序中对该类的使用。GameView.java 文件头如下：

```java
package com.as.minigame;                          //文件所属包

import java.util.ArrayList;                       //导入该类定义所需的其他类和函数
import java.util.List;
import java.util.Random;
import android.app.AlertDialog;
import android.content.Context;
import android.content.DialogInterface;
import android.graphics.Point;
import android.util.AttributeSet;
import android.view.MotionEvent;
import android.view.View;
import android.widget.GridLayout;
```

第 14 章　图形化数字游戏

2) 实现 GameView.java

GameView.java 的初始化单元如下：

```java
public class GameView extends GridLayout {

    private Card cards[][] = new Card[4][4];
    private List<Point> emptyCards = new ArrayList<Point>();
    Random rd = new Random();
    int score = 0;

    public GameView(Context context) {
        super(context);
        //TODO Auto-generated constructor stub
        initGame();
    }

    public GameView(Context context, AttributeSet attrs) {
        super(context, attrs);
        //TODO Auto-generated constructor stub
        initGame();
    }

    public GameView(Context context, AttributeSet attrs, int defStyle) {
        super(context, attrs, defStyle);
        //TODO Auto-generated constructor stub
        initGame();
    }
}
```

3) 游戏初始化函数

该函数的主要功能为设定游戏的输入监听，并定义如何判断用户的游戏输入，关联到相应的处理函数，具体代码如下：

```java
private void initGame() {
    setColumnCount(4);
    setBackgroundColor(0xffffcccc);

    setOnTouchListener(new OnTouchListener() {

        private float startX, startY;
        private float offsetX, offsetY;

        @Override
        public boolean onTouch(View v, MotionEvent event) {
            //TODO Auto-generated method stub

            switch (event.getAction()) {
                case MotionEvent.ACTION_DOWN:
                    startX = event.getX();
```

```
                    startY = event.getY();
                    break;
                case MotionEvent.ACTION_UP:

                    Gameover();

                    offsetX = event.getX() - startX;
                    offsetY = event.getY() - startY;
                    if (Math.abs(offsetX) > Math.abs(offsetY)) {
                        if (offsetX < -3) {
                            moveLeft();
                            System.out.println("----左");
                        } else if (offsetX > 3) {
                            moveRight();
                            System.out.println("----右");
                        }

                    } else {
                        if (offsetY < -3) {
                            moveUp();
                            System.out.println("----上");
                        } else if (offsetY > 3) {
                            moveDown();
                            System.out.println("----下");
                        }

                    }

                    break;
                default:
                    break;
            }

            return true;
        }
    });
}
```

4) 游戏触摸响应函数

该工程一共定义了上下左右四个触摸响应函数，该函数的主要功能为响应用户的输入，将结果数组按照输入的方向进行更新并创建新的随机数卡片，具体代码如下：

```
    private void moveRight() {

        boolean flage = false;
        for (int y = 0; y < 4; y++) {
            for (int x = 3; x >= 0; x--) {
                for (int x1 = x - 1; x1 >= 0; x1--) {
                    //当同一行为空时，不需处理
                    if (cards[x1][y].getNumber() > 0) {
```

第 14 章 图形化数字游戏

```java
            if (cards[x][y].getNumber() < 2) {
                //将前一张卡片的值移动到当前卡片
                cards[x][y].setNumber(cards[x1][y].getNumber());
                cards[x1][y].setNumber(0);
                x++;
                flage = true;
                score +=2;
            } else if (cards[x][y].getNumber() ==
                    cards[x1][y].getNumber()) {
                cards[x][y].setNumber(cards[x][y].getNumber() * 2);
                score +=cards[x][y].getNumber();
                cards[x1][y].setNumber(0);
                flage = true;
            }
            break;
        }
        }
    }
    }

    if (flage) {
        creatRandomCard();
    }
}

private void moveLeft() {

    boolean flage = false;
    for (int y = 0; y < 4; y++) {
        for (int x = 0; x < 4; x++) {
            for (int x1 = x + 1; x1 < 4; x1++) {
                //当同一行为空时，不需处理
                if (cards[x1][y].getNumber() > 0) {
                    if (cards[x][y].getNumber() < 2) {
                        //将前一张卡片的值移动到当前卡片
                        cards[x][y].setNumber(cards[x1][y].getNumber());
                            cards[x1][y].setNumber(0);
                        x--;
                        flage = true;
                        score +=2;
                    } else if (cards[x][y].getNumber() == cards[x1][y]
                            .getNumber()) {
                        cards[x][y].setNumber(cards[x][y].getNumber() * 2);
                        score +=cards[x][y].getNumber();
                        cards[x1][y].setNumber(0);
                        flage = true;
                    }
                    break;
                }
```

```java
            }
        }
    }
    if (flage) {
        creatRandomCard();
    }
}

private void moveDown() {
    boolean flage = false;
    for (int x = 0; x < 4; x++) {
        for (int y = 3; y >= 0; y--) {
            for (int y1 = y - 1; y1 >= 0; y1--) {
                //当同一行为空时,不需处理
                if (cards[x][y1].getNumber() > 0) {
                    if (cards[x][y].getNumber() < 2) {
                        //将前一张卡片的值移动到当前卡片
                        cards[x][y].setNumber(cards[x][y1].getNumber());
                        cards[x][y1].setNumber(0);
                        y++;
                        flage = true;
                        score +=2;
                    } else if (cards[x][y].getNumber() ==
                            cards[x][y1].getNumber()) {
                        cards[x][y].setNumber(cards[x][y].getNumber() * 2);
                        score +=cards[x][y].getNumber();
                        cards[x][y1].setNumber(0);
                        flage = true;
                    }
                    break;
                }
            }
        }
    }

    if (flage) {
        creatRandomCard();
    }
}

private void moveUp() {
    boolean flage = false;
    for (int x = 0; x < 4; x++) {
        for (int y = 0; y < 4; y++) {
            for (int y1 = y + 1; y1 < 4; y1++) {
                if (cards[x][y1].getNumber() > 0) {
                    if (cards[x][y].getNumber() < 2) {
                        cards[x][y].setNumber(cards[x][y1].getNumber());
```

第14章 图形化数字游戏

```
                cards[x][y1].setNumber(0);
                y--;
                flage = true;
                score+=2;
            } else if (cards[x][y].getNumber() == cards[x][y1]
                    .getNumber()) {
                cards[x][y].setNumber(cards[x][y].getNumber() * 2);
                score +=cards[x][y].getNumber();
                cards[x][y1].setNumber(0);
                flage = true;
            }
            break;
        }
    }
  }
}

if (flage) {
    creatRandomCard();
}
```

5) 游戏结束判定及处理函数

根据游戏设定的规则，如果所有卡片皆不为零，而且临近卡片分数都不相同不能合并的情况下，判断为游戏结束。游戏结束时，弹出对话框询问是否开始新游戏，具体代码如下：

```
private void Gameover(){
    boolean OverGame=true;
    for (int y = 0; y < 4; y++) {
        for (int x = 0; x < 4; x++) {
            if(cards[x][y].getNumber() <= 0 ||
                (x >0 && cards[x][y].getNumber() ==
                    cards[x-1][y].getNumber())||
                (x<3 && cards[x][y].getNumber() ==
                    cards[x+1][y].getNumber()) ||
                (y>0 && cards[x][y].getNumber() ==
                    cards[x][y-1].getNumber()) ||
                (y<3 && cards[x][y].getNumber() ==
                    cards[x][y+1].getNumber())
            ){
                OverGame=false;
            }
        }
    }
    if(OverGame){
        new AlertDialog.Builder(getContext()).setTitle("hi")
            .setMessage("again").
            setPositiveButton("yes",new AlertDialog.OnClickListener() {
            @Override
```

```java
                public void onClick(DialogInterface dialog, int which) {
                    //TODO Auto-generated method stub
                    GameStart();
                    score = 0;
                }
            }).setNegativeButton("No", null).show();
    }
}
```

6) 其他函数

为满足程序需要，还需要定义其他函数来满足程序的功能。其中包括卡片的添加函数"AddCard(int width, int height)"；界面尺寸变化响应函数（当创建时也会调用）"onSizeChanged(int w, int h, int oldw, int oldh)"；创建随机有分卡片的"creatRandomCard()"函数；游戏数组初始化的开始函数"GameStart()"。具体代码如下：

```java
    private void AddCard(int width, int height) {
        Card c;
        for (int y = 0; y < 4; y++) {
            for (int x = 0; x < 4; x++) {
                c = new Card(getContext());
                cards[x][y] = c;
                c.setNumber(0);
                addView(c, width, height);
            }
        }
    }

    @Override
    protected void onSizeChanged(int w, int h, int oldw, int oldh) {
        //TODO Auto-generated method stub
        super.onSizeChanged(w, h, oldw, oldh);
        int width = (w - 10) / 4;
        AddCard(width, width);
        GameStart();
    }

    private void creatRandomCard() {
        emptyCards.clear();
        for (int y = 0; y < 4; y++) {
            for (int x = 0; x < 4; x++) {
                if (cards[x][y].getNumber() < 2) {
                    Point point = new Point(x, y);
                    emptyCards.add(point);
                }
            }
        }
        int selat = rd.nextInt(emptyCards.size());
        Point p = emptyCards.get(selat);
```

```
        emptyCards.remove(selat);
        int number = 0;
        if (rd.nextInt(10) > 4) {
            number = 4;
        } else
            number = 2;
        cards[p.x][p.y].setNumber(number);
    }

    public void GameStart() {
        for (int y = 0; y < 4; y++) {
            for (int x = 0; x < 4; x++) {
                cards[x][y].setNumber(0);
            }
        }
        creatRandomCard();
        creatRandomCard();
    }
```

14.4 系统运行与测试

界面设计和代码设计完成后,单击 Android Studio 开发环境的工具栏中的 ▶ 图标,或者选择菜单栏中的"Run"→"Run"命令,运行该项目,可以调试查看 Android 数字小游戏程序。单击运行后,会弹出设备选择窗口,如图 14.11 所示。这里可以选择连接在计算机上的手机设备、运行中的 Android 虚拟机或者新启动一个 Android 虚拟设备。这里选择启动早先使用 AVD 创建的"Nexus S"设备。

图 14.11 设备选择界面

游戏运行效果如图 14.12 所示。

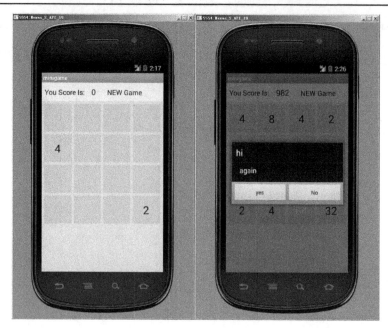

图 14.12 游戏运行效果——游戏开始及结束

第 15 章　基于 Android 的音频播放器

15.1　系统分析

手机中的音频播放器是一种在日常生活中放松身心或活跃气氛的常用休闲软件。无论是在日常休闲还是旅游出差途中，或是家庭娱乐中都有着它的身影。随着移动互联网和智能手机的不断发展和进步，当今市场上也出现了多种音频播放软件。本章介绍怎样制作一款简单实用的音频播放软件。

（1）要求

播放器的功能要符合使用者的基本情况，可以方便地进行操作，播放器的界面要求美观，有自己的特色。系统的界面要简易、快捷，不要有过多的修饰或者不必要的功能；软件要有一定的安全性和健壮性，不能影响手机系统的运行。

（2）目标

方便所有对音乐歌曲欣赏功能有需求的手机用户。

15.2　系统设计

15.2.1　系统目标

根据用户对 Android 音频播放器的使用要求，制定如下系统目标。
- 操作简单、易于掌握，界面简洁清爽。
- 方便的音乐歌曲的播放以及播放控制功能。
- 可方便地对乐曲库进行增、删、改管理。
- 系统运行稳定，不能和手机固有软件冲突，安全可靠。

15.2.2　系统功能结构

Android 音频播放器的功能结构图如图 15.1 所示。

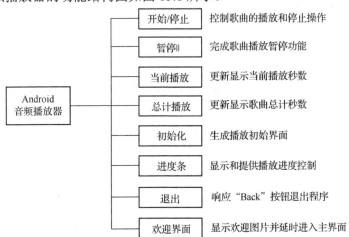

图 15.1　Android 音频播放器功能结构图

15.2.3 系统业务流程

Android 音频播放器软件的业务流程图如图 15.2 所示。

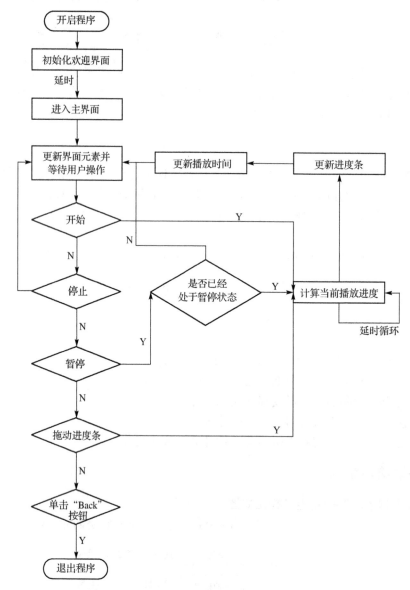

图 15.2 Android 音频播放器软件的业务流程图

15.3 系 统 实 施

15.3.1 开发及运行环境

本项目的软件开发及运行环境如下。
- 操作系统：Windows 7。

- 开发工具：Android Studio 1.1.0 + Android 4.4.2（API 19）。
- JDK 环境：Java SE Development Kit（JDK）1.8.0_31。
- 开发语言：Java、XML。
- 运行平台：AVD Nexus S API 19（虚拟机设置）。
- 分辨率：480×800 hdpi。

15.3.2 项目的创建

Android 音频播放器系统的项目名称为 MiniPlayer，该项目是使用 Android Studio 1.1.0 + Android 4.4.2（API 19）开发的，在 Android Studio 开发环境中创建该项目的步骤如下。

（1）启动 Android Studio，经过初始化配置之后，出现如图 15.3 所示的界面。在这个界面中可以新建项目，也可以导入本地项目等，在左侧可以查看最近打开的项目等，这里选择新建项目。

图 15.3　初始化窗口

（2）单击"Start a new Android Studio project"选项后，会弹出 Create New Project 窗口。在该窗口中首先输入项目名称 MiniPlayer，并且输入公司域名为 as.com，然后输入项目的存储路径（如图 15.4 所示），最后单击"Next"按钮。

（3）弹出的"Target Android Devices"页面支持用户适配 TV、Wear、Glass 等，这里只选择第一项 Phone and Tablet，然后选好项目支持的最小 SDK（API 19），最后单击"Next"按钮，如图 15.5 所示。

（4）在弹出的"Add Activity"页面中选择一个 Activity 模板添加到工程中，和 Eclipse 很像，这里直接选择"Blank Activity"创建一个"Hello World！"程序，然后单击"Next"按钮，如图 15.6 所示。

（5）在弹出的"Customize the Activity"页面中可以设置主 Activity 的名称、XML 界面配置文件名称、项目的 Title 和菜单名称，然后单击"Finish"按钮完成工程的创建，如图 15.7 所示。

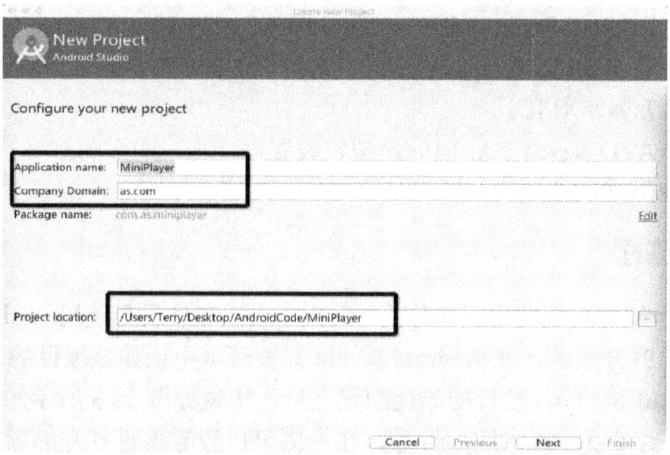

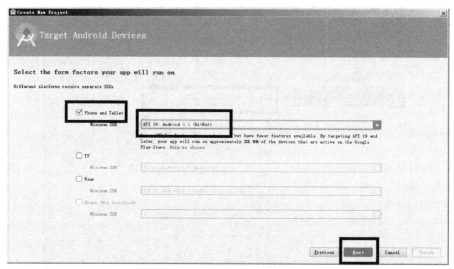

图 15.4 输入项目名称和路径等

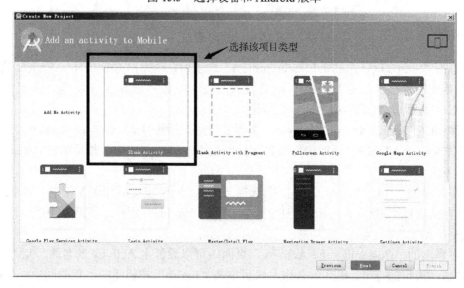

图 15.5 选择设备和 Android 版本

图 15.6 添加 Activity

第 15 章 基于 Android 的音频播放器

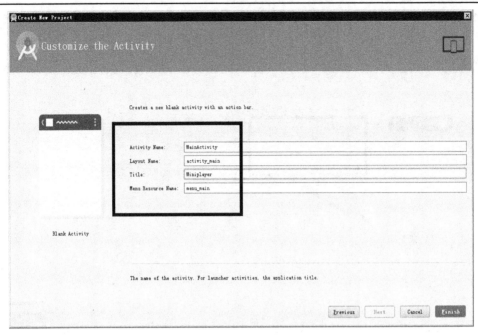

图 15.7 设置文件名和 Title 等

(6) 由于本项目设计有多个 Activity，于是在创建好工程和主 Activity 之后，还需要创建其他新的 Activity。创建新的 Activity 需要在项目窗口中右键单击"Android.app.res.layout"图标（如图 15.8 所示），选择"New→Activity→Black Activity"（或者从菜单栏中选择相应菜单项创建）。

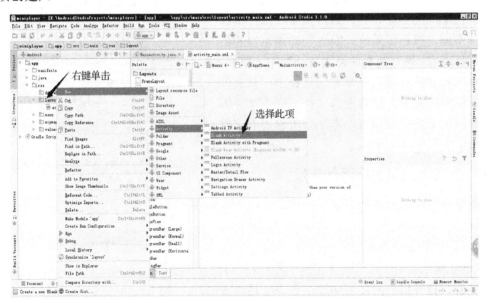

图 15.8 创建新的 Activity

(7) 在出现的界面中，设置好新的 Activity 的各种属性名字和显示的 Title，由于这个界面将用于显示欢迎界面，所以勾选设置为启动 Activity（如图 15.9 所示）。

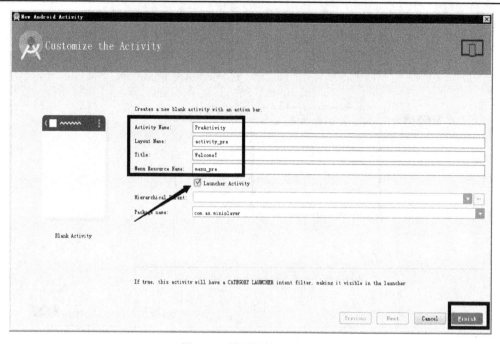

图 15.9 设置新的 Activity

15.3.3 项目工程结构

在编写代码之前，首先要制定好项目的系统文件夹的组织结构目录，分门别类地管理我们的项目资源，包括不同的界面、类、数据模型、图片资源等，这样不但可以保证系统开发过程的规范性和程序员的可替代性，还有利于保证团队开发的一致性。创建好系统的文件夹后，在开发的过程中，只需要将我们新创建的类文件或者资源文件、脚本文件等保存到相应的文件夹中即可。本项目所使用的文件夹组织结构如图 15.10 所示。

15.3.4 音频播放器主界面的实现

界面是程序必不可少的元素，是实现人机交互的必有环节。通过界面，用户可以链接到系统的各个子模块，快速地调用系统实现的所有功能。在 Android 音频播放器中，我们设计了两个界面：一个是欢迎界面，提升用户的使用体验；另一个是播放界面，实现歌曲播放功能。欢迎界面显示音乐图标并自动过渡到播放界面，在播放界面中实现播放、快进、后退、暂停等功能。界面运行结果如图 15.11 所示。

图 15.10 文件夹组织结构

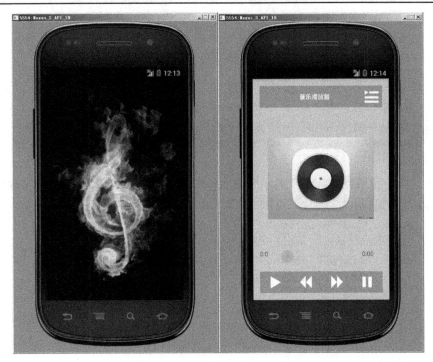

图 15.11　Android 音频播放器界面

1. 欢迎界面布局 XML 文件

本次工程拥有两个界面，并实现自动切换，我们分别来解读这两个界面的布局文件。首先是欢迎界面，主要显示欢迎图片，所以包含一个图片控件。具体代码如下：

```xml
<?xml version="1.0" encoding="utf-8"?>
<LinearLayout xmlns:android="http://schemas.android.com/apk/res/android"
    android:layout_width="match_parent"
    android:layout_height="match_parent"
    android:background="#000000"
    android:orientation="vertical" >

    <ImageView
        android:layout_width="match_parent"
        android:layout_height="match_parent"
        android:src="@drawable/de" />

</LinearLayout>
```

2. 播放界面布局 XML 文件

为实现歌曲播放的可控制性和友好性，播放界面的设计比较复杂，一共设计了 4 个子界面(布局管理器)来完成整个界面的布局。

具体的布局代码如下：

```xml
<LinearLayout xmlns:android="http://schemas.android.com/apk/res/android"
    xmlns:tools="http://schemas.android.com/tools"
    android:layout_width="match_parent"
```

```
        android:layout_height="match_parent"
        android:background="#c0c0c0"
        android:orientation="vertical"
        android:paddingBottom="@dimen/activity_vertical_margin"
        android:paddingLeft="@dimen/activity_horizontal_margin"
        android:paddingRight="@dimen/activity_horizontal_margin"
        android:paddingTop="@dimen/activity_vertical_margin"
        tools:context=".MainActivity"
</LinearLayout>
```

第一个界面：顶层界面，用于显示歌曲名称和显示歌曲列表按钮(功能待实现)。

```
<RelativeLayout
    android:id="@+id/one"
    android:layout_width="match_parent"
    android:layout_height="@dimen/b"
    android:background="#808a87" >

    <TextView
        android:id="@+id/musicname"
        android:layout_width="@dimen/c"
        android:layout_height="wrap_content"
        android:layout_centerInParent="true"
        android:text="@string/yb"
        android:textColor="#ffffff" />

    <Button
        android:id="@+id/meun"
        android:layout_width="@dimen/b"
        android:layout_height="@dimen/b"
        android:layout_alignParentRight="true"
        android:background="@drawable/meun" />
</RelativeLayout>
```

第二个子界面：中部界面，用于显示歌曲图片。

```
<RelativeLayout
    android:id="@+id/images"
    android:layout_width="match_parent"
    android:layout_height="@dimen/a"
    android:layout_weight="1" >

    <ImageView
        android:id="@+id/image"
        android:layout_width="@dimen/f"
        android:layout_height="@dimen/f"
        android:layout_centerInParent="true"
        android:src="@drawable/maiyatang" />
</RelativeLayout>
```

第三子个界面：播放进度界面，用于显示播放进度条和播放时长与歌曲总时长。

```xml
<LinearLayout
    android:layout_width="match_parent"
    android:layout_height="@dimen/b" >

    <TextView
        android:id="@+id/starttime"
        android:layout_width="@dimen/a"
        android:layout_height="match_parent"
        android:layout_weight="1"
        android:text="@string/t1"
        android:textColor="#ff0000" />

    <SeekBar
        android:id="@+id/seekbar"
        android:layout_width="@dimen/a"
        android:layout_height="wrap_content"
        android:layout_weight="4" />

    <TextView
        android:id="@+id/overtime"
        android:layout_width="@dimen/a"
        android:layout_height="match_parent"
        android:layout_weight="1"
        android:text="@string/t2"
        android:textColor="#ff0000" />
</LinearLayout>
```

第四个子界面：底部按钮界面，用于显示播放控制的四个按钮，这四个按钮又分别以二层子界面的形式来定义制作。

```xml
<LinearLayout
    android:id="@+id/bottom"
    android:layout_width="match_parent"
    android:layout_height="@dimen/b"
    android:background="#808a87" >

    <RelativeLayout
        android:id="@+id/b1"
        android:layout_width="@dimen/a"
        android:layout_height="@dimen/b"
        android:layout_weight="1" >

        <Button
            android:id="@+id/stop"
            android:layout_width="@dimen/d"
            android:layout_height="@dimen/d"
            android:layout_centerInParent="true"
```

```xml
            android:background="@drawable/stop" />
    </RelativeLayout>

    <RelativeLayout
        android:id="@+id/b2"
        android:layout_width="@dimen/a"
        android:layout_height="@dimen/b"
        android:layout_weight="1" >

        <Button
            android:id="@+id/left"
            android:layout_width="@dimen/b"
            android:layout_height="@dimen/b"
            android:layout_centerInParent="true"
            android:background="@drawable/left" />
    </RelativeLayout>

    <RelativeLayout
        android:id="@+id/b3"
        android:layout_width="@dimen/a"
        android:layout_height="@dimen/b"
        android:layout_weight="1" >

        <Button
            android:id="@+id/right"
            android:layout_width="@dimen/b"
            android:layout_height="@dimen/b"
            android:layout_centerInParent="true"
            android:background="@drawable/right" />
    </RelativeLayout>

    <RelativeLayout
        android:id="@+id/b4"
        android:layout_width="@dimen/a"
        android:layout_height="@dimen/b"
        android:layout_weight="1" >

        <Button
            android:id="@+id/swit"
            android:layout_width="@dimen/d"
            android:layout_height="@dimen/d"
            android:layout_centerInParent="true"
            android:background="@drawable/swit" />
    </RelativeLayout>
</LinearLayout>
```

15.3.5 Android 音频播放器逻辑的实现

由于 Android 音频播放器需要一定的加强用户体验，所以我们设计了两个 activity 来过渡显示。另外，也设计了两个 java 类文件来完成所有程序功能，其中 PreActivity 用于显示欢迎界面并跳转到主界面，MainActivity 用于实现主界面的所有功能。建立工程后，除默认的主程序文件，也就是Android Studio工程界面中的java文件夹内的"com.as.miniplayer"包中的"MainActivity.java"文件外，再创建一个"Java Class"，系统会自动生成与类名同名的文件"PreActivity.java"。文件所在位置如图 15.12 所示。下面将对这两个文件分模块解析说明。

1. PreActivity.java 文件

顾名思义，Android 音频播放器工程中的"PreActivity.java"文件主要包含了欢迎界面类的定义，以及各个元件的定义和指定界面元素与响应函数的关联，并包含自动跳转到主界面的功能，下面详细解析该类的各个模块。

1) 文件导入的包和类

由于该类需要使用库函数中的模块定义，所以在编写程序时，需"import（导入）"相应的内容来支持程序中对该类的使用。例如，要使用Android Studio 中的"文本"元件，就必须导入"TextView"类，Android 音频播放器"PreActivity.java"的文件头如下：

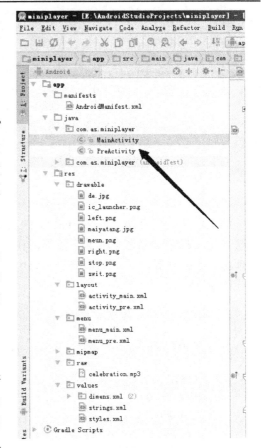

图 15.12 Android 源程序文件

```
package com.as.miniplayer;

import android.app.Activity;
import android.content.Intent;
import android.os.Bundle;
import android.os.Handler;
import android.os.Message;
```

2) 类的定义

"PreActivity.java"的初始化单元如下：

```
public class PreActivity extends Activity{
    Handler handler=new Handler(){
        public void handleMessage(Message msg) {
            if(msg.what==250){
                Intent intent=new Intent(PreActivity.this,MainActivity.class);
                startActivity(intent);             //执行跳转页面方法
                finish();                          //销毁当前页面
            }
        }
    };
```

```
    };
    @Override
    protected void onCreate(Bundle savedInstanceState) {
        //TODO Auto-generated method stub
        super.onCreate(savedInstanceState);
        setContentView(R.layout.activity_pre);      //加载布局
        WaitThread thread=new WaitThread();         //开启线程
        thread.start();
    }
```

3) 定义线程

该工程从欢迎界面 "PreActivity" 跳转到主界面 "MainActivity" 是采用的线程间通信的方法，先发起一个包含延时的子线程，在延时结束后由子线程给主线程发消息，然后在消息处理函数中跳转界面。上一小节的代码中已经包含了消息判断和跳转的内容，定义线程的详细代码如下：

```
/**
 *
 * 暂留5s跳转
 *
 * */
public class WaitThread extends Thread{//创建线程
    @Override
    public void run() {
        //TODO Auto-generated method stub
        try {
            Thread.sleep(5000);
        } catch (InterruptedException e) {
            // TODO Auto-generated catch block
            e.printStackTrace();
        }
        //写一封信
        Message message =new Message();
        message.what=250;
        message.arg1=1;
        handler.sendMessage(message);//发送
    }
}
```

2. 实现 MainActivity.java 文件

1) 文件导入的包和类

由于该类需要使用库函数中的模块定义，所以在编写程序时，需 "import" 相应的内容来支持程序中对该类的使用。MainActivity.java 文件头如下：

```
package com.as.miniplayer;

import java.io.IOException;
```

第15章 基于Android的音频播放器

```java
import android.media.MediaPlayer;
import android.os.Bundle;
import android.os.Handler;
import android.app.Activity;
import android.view.KeyEvent;
import android.view.View;
import android.view.View.OnClickListener;
import android.widget.Button;
import android.widget.ImageView;
import android.widget.SeekBar;
import android.widget.SeekBar.OnSeekBarChangeListener;
import android.widget.TextView;
```

2) 类的定义

"MainActivity.java" 的初始化单元如下：

```java
/***implements 是实现监听***/
public class MainActivity extends Activity implements OnClickListener {
    private TextView musicName;        //上边显示歌名
    private Button menu;               //菜单按钮
    private ImageView image;           //显示图片
    private TextView startTime;
    private TextView countTime;
    private SeekBar seekBar;
    private Button play;
    private Button stop;
    private Button next;
    private Button last;
    private MediaPlayer player;
    @Override
    protected void onCreate(Bundle savedInstanceState) {
        super.onCreate(savedInstanceState);
        setContentView(R.layout.activity_main);
        initView();
        player=MediaPlayer.create(this, R.raw.celebration);//初始化音乐播放器
        seekBar.setMax(player.getDuration());              //设最大值
        seekBar.setOnSeekBarChangeListener(new OnSeekBarChangeListener() {

            @Override
            public void onStopTrackingTouch(SeekBar seekBar) {
                //TODO Auto-generated method stub

            }

            @Override
            public void onStartTrackingTouch(SeekBar seekBar) {
                //TODO Auto-generated method stub
```

```java
            }

            @Override
            public void onProgressChanged(SeekBar seekBar, int progress,
                                boolean fromUser) {
                //TODO Auto-generated method stub
                if(fromUser){
                    player.seekTo(progress);
                }
            }
        });
    }
```

3)初始化

该函数的主要功能为设定界面显示控件与参数的关系，进行按钮的输入监听，关联到相应的处理函数，具体代码如下：

```java
    private void initView() {
        //TODO Auto-generated method stub
        musicName=(TextView) findViewById(R.id.musicname);
        menu=(Button) findViewById(R.id.meun);
        image=(ImageView) findViewById(R.id.image);
        startTime=(TextView) findViewById(R.id.starttime);
        countTime=(TextView) findViewById(R.id.overtime);
        seekBar=(SeekBar) findViewById(R.id.seekbar);
        play=(Button) findViewById(R.id.stop);
        stop=(Button) findViewById(R.id.swit);
        next=(Button) findViewById(R.id.right);
        last=(Button) findViewById(R.id.left);
        menu.setOnClickListener(this);
        play.setOnClickListener(this);
        stop.setOnClickListener(this);
        next.setOnClickListener(this);
        last.setOnClickListener(this);
    }
```

4)播放控制触摸响应函数

该工程一共定义了"播放列表"、"播放/停止"、"暂停"、"上一首"、"下一首"五个触摸按钮响应函数，其中实现了"播放/停止"、"暂停"两个按钮的功能，对其余未实现功能有兴趣的读者可自行开发，具体代码如下：

```java
    @Override
    public void onClick(View v) {
        //TODO Auto-generated method stub
        int id=v.getId();//单击按钮的id
        switch (id) {
            case R.id.meun:
```

```
                break;
            case R.id.stop:
                player.start();
                handler.post(run);//开启
                break;
            case R.id.swit:
                player.stop();
                try {
                    player.prepare();
                } catch (IllegalStateException e) {
                    //TODO Auto-generated catch block
                    e.printStackTrace();
                } catch (IOException e) {
                    //TODO Auto-generated catch block
                    e.printStackTrace();
                }
                break;
            case R.id.right:

                break;
            case R.id.left:

                break;
        }
    }
```

5) 界面显示元素的更新函数

根据本项目界面的设定，需要在播放的同时，更新进度条进度、目前播放时间、总播放时间三个界面元素。这里通过子线程 "run()" 来实现，具体代码如下：

```
    Handler handler=new Handler();
    Runnable run= new Runnable() {

        @Override
        public void run() {
            seekBar.setProgress(player.getCurrentPosition());    //音乐播放器当前值
            setTime();
            handler.postDelayed(run, 200);                        //延时调用
        }
    };
    public void setTime(){                                        //设置时间的方法
        int now =player.getCurrentPosition();                     //获取当前播放速度
        int count =player.getDuration();                          //获取总进度
        int second = now/1000;                                    //当前有多少秒
        int csecond =count/1000;                                  //总共有多少秒
        startTime.setText(second/60+":"+second%60);
        countTime.setText(csecond/60+":"+csecond%60);
    }
```

6)退出函数

为满足程序需要,还需要定义退出函数来退出整个程序的运行,同时做好程序的资源清理工作,包括清理线程、释放资源等。具体代码如下(注意退出的先后顺序):

```
@Override      //按钮监听事件,keyCode 是按下按钮的编码
public boolean onKeyDown(int keyCode, KeyEvent event) {
    //TODO Auto-generated method stub
    if(keyCode==KeyEvent.KEYCODE_BACK){
        player.stop();
        player.release();                    //释放资源
        handler.removeCallbacks(run);        //线程移除
        finish();}
    return super.onKeyDown(keyCode, event);
}
```

15.4 系统运行与测试

界面设计和代码设计完成后,单击 Android Studio 开发环境的工具栏中的 ▶ 图标,或者选择菜单栏中的"Run"→"Run"命令,运行该项目,可以调试查看 Android 音频播放器程序。当单击运行后,会弹出设备选择窗口,如图 15.13 所示,这里可以选择连接在计算机上的手机设备、运行中的 Android 虚拟机或者新启动一个 Android 虚拟设备。我们选择启动早先用 AVD 创建好的"Nexus S"设备。

在运行好的窗口中单击"播放"按钮播放内置音乐,播放成功,项目开发完成!如图 15.14 所示。

图 15.13 设备选择窗口   图 15.14 项目运行结果——播放音乐的过程

第16章 Android 数据库记事本

16.1 系统分析

记事本是一种在日常生活中应用广泛的应用。无论是学生、上班族还是自由职业者，常常需要处理较多的事情，事情一旦繁杂起来就比较容易混乱或者忘记。有时候，记录事情的纸张可能会不小心弄丢，而且不方便随时随地记录和查看。本章介绍的 Android 数据库记事本采用数据库存储数据，让读者更加了解 SQLite 的使用方法。

（1）要求

软件的功能要符合使用者的基本情况；可以方便地进行事件记录、查看、更改、删除等操作；系统的界面要简易、快捷，不要有过多的修饰或者不必要的功能；软件运行流程，不能影响手机系统的运行。

（2）目标

方便所有对记事本功能有需求的手机用户。

16.2 系统设计

16.2.1 系统目标

根据用户对 Android 数据库记事本的使用要求，制定系统目标如下：
- 操作简单、易于掌握，界面简洁清爽。
- 方便进行记录、查看、删除、修改等操作。
- 包含输入和返回功能。
- 系统运行稳定，不能和手机固有软件冲突，安全可靠。

16.2.2 系统功能结构

Android 数据库记事本的功能结构图如图 16.1 所示。

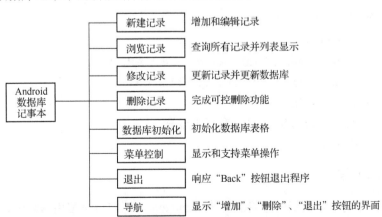

图 16.1 Android 数据库记事本的功能结构图

16.2.3 系统业务流程

Android 数据库记事本软件的业务流程图如图 16.2 所示。

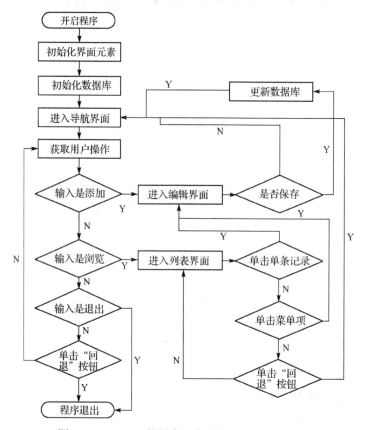

图 16.2 Android 数据库记事本软件的业务流程图

16.3 系统实施

16.3.1 开发及运行环境

本项目的软件开发及运行环境如下。
- 操作系统：Windows 7。
- 开发工具：Android Studio 1.1.0 + Android 4.4.2（API 19）。
- JDK 环境：Java SE Development Kit（JDK）1.8.0_31。
- 开发语言：Java、XML。
- 运行平台：AVD Nexus S API 19（虚拟机设置）。
- 分辨率：480×800 hdpi。

16.3.2 项目的创建

Android 数据库记事本系统的项目名称为 MiniNote，该项目是使用 Android Studio 1.1.0 + Android 4.4.2（API 19）开发的，在 Android Studio 开发环境中创建该项目的步骤如下。

(1)启动 Android Studio,经过初始化配置之后,会出现如图 16.3 所示的界面。在这个界面中可以新建项目,也可以导入本地项目等,在左侧可以查看最近打开的项目等,这里选择新建项目。

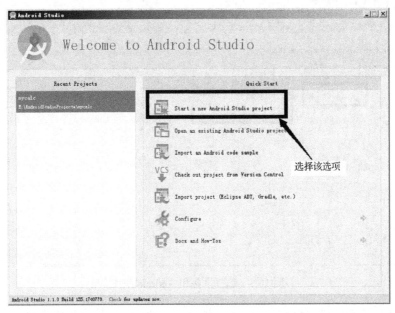

图 16.3 初始化窗口

(2)单击"Start a new Android Studio Project"选项后,会弹出 Create New Project 窗口。在该窗口中首先输入项目名称 MiniNote,并且输入公司域名为 as.com,然后输入项目的存储路径(如图 16.4 所示),最后单击"Next"按钮。

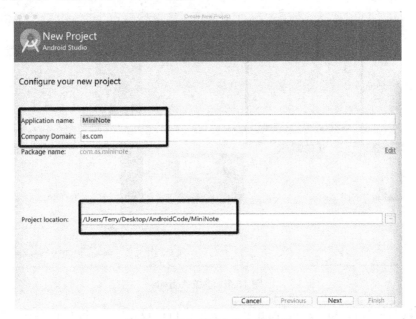

图 16.4 输入项目名称和路径等

(3)弹出的"Target Android Devices"页面支持用户适配 TV、Wear、Glass 等,这里只选

择第一项"Phone and Table",然后选好项目支持的最小 SDK(API 19),最后单击"Next"按钮,如图 16.5 所示。

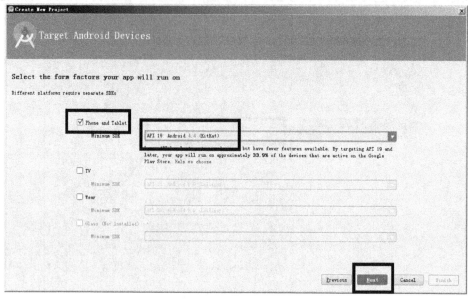

图 16.5　选择设备和 Android 版本

(4)在弹出的"Add Activity"页面中选择一个 Activity 模板添加到工程中,和 Eclipse 很像,直接选择"Blank Activity"创建一个"Hello World!"程序,然后单击"Next"按钮,如图 16.6 所示。

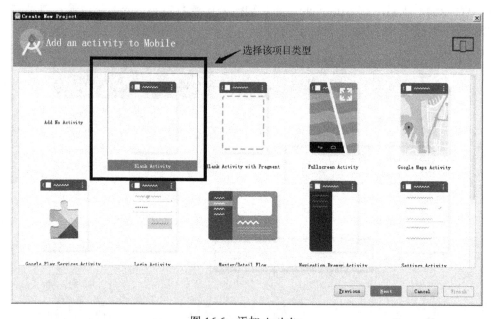

图 16.6　添加 Activity

(5)在弹出的"Customize the Activity"页面中可以设置主 Activity 的名称、XML 界面配置文件名称、项目的 Title 和菜单名称,然后单击"Finish"按钮完成工程的创建,如图 16.7 所示。

第 16 章 Android 数据库记事本

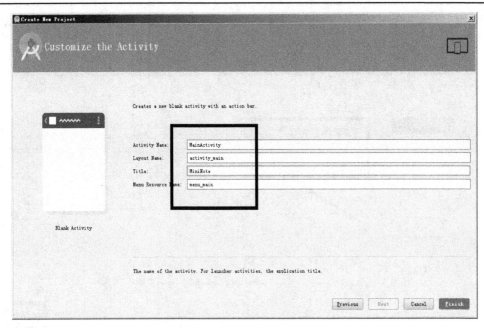

图 16.7 设置文件名和 Title

(6) 由于本项目设计有多个 Activity，所以在创建好工程和主 Activity 之后，还需要创建导航 Activity、列表 Activity。创建新的 Activity 需要在项目窗口中右键单击"Android.app.res.layout"图标（如图 16.8 所示），选择"New→Activity→Black Activity"（或者从菜单栏中选择相应的菜单项创建）。

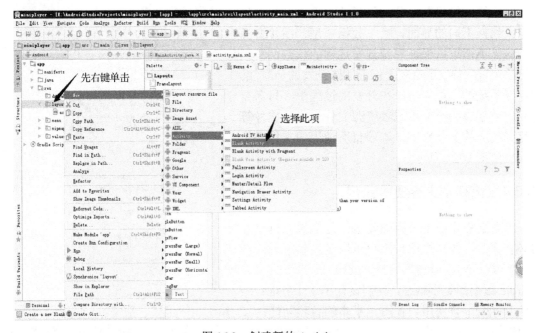

图 16.8 创建新的 Activity

(7) 在出现的界面中，设置新的 Activity 的各种属性名字和显示的 Title，如果需要将此界面设置为启动界面，则勾选设置为启动 Activity（如图 16.9 所示）。

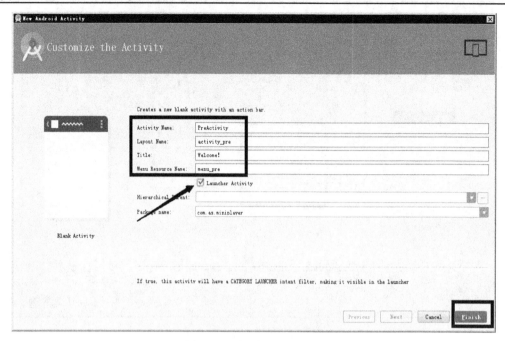

图 16.9　设置新的 Activity

16.3.3　项目工程结构

在编写代码之前，首先要制定好项目的系统文件夹的组织结构目录，分门别类地管理我们的项目资源，包括不同的界面、类、数据模型、图片资源等，这样不但可以保证系统开发过程的规范性和程序员的可替代性，还有利于保证团队开发的一致性。创建好系统的文件夹后，在开发的过程中，只需要将我们新创建的类文件或者资源文件、脚本文件等保存到相应的文件夹中。本项目所使用的文件夹组织结构如图 16.10 所示。

16.3.4　记事本主界面的实现

界面是程序必不可少的元素，是实现人机交互的必有环节。通过界面，用户可以链接到系统的各个子模块，快速地调用系统实现的所有功能。在 Android 数据库记事本中，我们设计了 3 个界面：一是导航界面，方便用户使用记事本的各项功能；二是列表界面，方便用户浏览和查看所有的记事列表；三是编辑界面，用于增加、修改、删除各项记事内容。界面运行结果如图 16.11 所示。

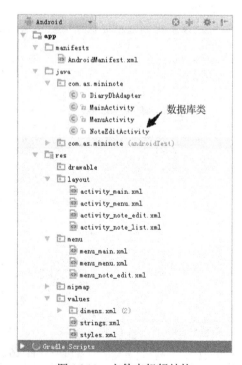

图 16.10　文件夹组织结构

第16章 Android 数据库记事本

图 16.11 Android 记事本界面

1. 导航界面布局 activity_memu.xml 文件实现

本次工程拥有三个界面,并实现互相可切换。下面分别解读这三个界面的布局文件。首先,是导航界面 activity_memu.xml 文件,主要显示切换到另外两个界面的按钮以及"退出"按钮,并包含一个欢迎语句的显示。具体代码如下:

```xml
<?xml version="1.0" encoding="utf-8"?>
<RelativeLayout xmlns:android="http://schemas.android.com/apk/res/android"
    android:layout_width="fill_parent"
    android:layout_height="fill_parent" >

    <TextView
        android:id="@+id/textView1"
        android:layout_width="wrap_content"
        android:layout_height="wrap_content"
        android:layout_alignParentTop="true"
        android:layout_centerHorizontal="true"
        android:layout_marginTop="39dp"
        android:text="@string/wel_title"
        android:textAppearance="?android:attr/textAppearanceLarge" />

    <Button
        android:id="@+id/all"
        android:layout_width="wrap_content"
        android:layout_height="wrap_content"
        android:layout_alignBaseline="@+id/newinsert"
        android:layout_alignBottom="@+id/newinsert"
        android:layout_alignParentLeft="true"
        android:layout_marginLeft="15dp"
        android:text="@string/all" />

    <Button
```

```xml
        android:id="@+id/exitp"
        android:layout_width="wrap_content"
        android:layout_height="wrap_content"
        android:layout_below="@+id/textView1"
        android:layout_marginLeft="25dp"
        android:layout_marginTop="134dp"
        android:layout_toRightOf="@+id/textView1"
        android:text="@string/exitp" />

    <Button
        android:id="@+id/newinsert"
        android:layout_width="wrap_content"
        android:layout_height="wrap_content"
        android:layout_alignBaseline="@+id/exitp"
        android:layout_alignBottom="@+id/exitp"
        android:layout_alignLeft="@+id/textView1"
        android:text="@string/newinsert" />

</RelativeLayout>
```

2. 编辑界面布局 activity_note_edit.xml 文件实现

为实现记事本内容的增、删、改功能，设计一个编辑界面来完成这三个功能。当需要增加时，打开空白的编辑界面，并把结果存入数据库；当需要修改时，打开已有内容的编辑界面，并可修改内容后把结果存入数据库；当需要删除时，同样打开该条内容的编辑界面，让用户看着内容确认删除，并更新数据库。

综上所述，增、删、改功能可以用同一个界面布局来实现。该界面由标题、内容、按钮三个子界面组成。具体的布局代码如下：

```xml
<?xml version="1.0" encoding="utf-8"?>
<LinearLayout xmlns:android="http://schemas.android.com/apk/res/android"
    android:orientation="vertical" android:layout_width="fill_parent"
    android:layout_height="fill_parent">

    <LinearLayout android:orientation="vertical"
        android:layout_width="fill_parent"
        android:layout_height="wrap_content">

        <TextView
            android:layout_width="wrap_content"
            android:layout_height="wrap_content"
            android:text="@string/title"
            android:padding="2dp" />
        <EditText
            android:id="@+id/title"
            android:hint="@string/hint_title"
            android:layout_width="fill_parent"
            android:layout_height="wrap_content"
```

```xml
            android:layout_weight="1" />
    </LinearLayout>

    <TextView android:layout_width="wrap_content"
        android:layout_height="wrap_content" android:text="@string/body" />
    <EditText
        android:id="@+id/body"
        android:hint="@string/hint_content"
        android:layout_width="fill_parent"
        android:layout_height="fill_parent"
        android:layout_weight="1"
        android:scrollbars="vertical"
        android:gravity="top" />

    <RelativeLayout
        android:layout_width="fill_parent"
        android:layout_height="wrap_content" >

        <Button
            android:id="@+id/confirm"
            android:layout_width="wrap_content"
            android:layout_height="wrap_content"
            android:layout_alignParentLeft="true"
            android:layout_alignParentTop="true"
            android:layout_marginLeft="84dp"
            android:text="@string/confirm" />

        <Button
            android:id="@+id/delet"
            android:layout_width="wrap_content"
            android:layout_height="wrap_content"
            android:layout_alignParentTop="true"
            android:layout_marginLeft="38dp"
            android:layout_toRightOf="@+id/confirm"
            android:text="@string/delet" />
    </RelativeLayout>
</LinearLayout>
```

3. 列表界面布局 activity_note_list.xml 文件

为便于用户整体查看和管理所有的记事条目，必须设计一个列表界面。并且，在列表界面中为每条记录显示其创建时间，更加有利于提升用户的管理方便性和使用体验。所以，该界面包含两个 TextView 显示记事标题和时间，具体代码如下：

```xml
<?xml version="1.0" encoding="utf-8"?>
<RelativeLayout android:id="@+id/row"
    xmlns:android="http://schemas.android.com/apk/res/android"
    android:layout_width="fill_parent"
    android:layout_height="fill_parent">
```

```xml
<TextView
    android:id="@+id/text1"
    android:layout_width="wrap_content"
    android:layout_height="100px"
    android:layout_alignParentLeft="true"
    android:layout_alignParentTop="true"
    android:maxWidth="200dip"
    android:text="第一组 第一项"
    android:textSize="24sp" />

<TextView
    android:id="@+id/created"
    android:layout_width="wrap_content"
    android:layout_height="75px"
    android:layout_alignBottom="@+id/text1"
    android:layout_alignParentRight="true"
    android:text="2016年1月3日" />
</RelativeLayout>
```

16.3.5 记事本逻辑的实现

由于 Android 数据库记事本需要一定的加强用户体验，所以我们设计了三个 activity 来完成所有功能。另外，也设计了三个 java 类文件来完成所有程序编写，其中 MenuActivity 用于显示导航按钮并提供退出程序的接口；MainActivity 用于实现记录列表的展示和管理；NoteEditActivity 用于单条记录的编写、修改、删除管理。建立工程后，除默认的主程序文件，即 Android Studio 工程界面中的 java 文件夹内的 "com.as.mininote" 包中的 "MainActivity.java" 文件外，再创建两个 "Java Class"，分别命名为 "MenuActivity.java" 和 "NoteEditActivity"。文件所在位置如图 16.12 所示，下面将对这三个文件分模块解析说明。

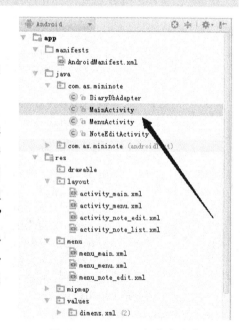

图 16.12　Android 源程序文件

1. 实现菜单 MenuActivity.java 文件

顾名思义，Android 数据库记事本工程中的 "MenuActivity.java" 文件主要包含了菜单（导航）界面类的定义，以及各个元件的定义和指定界面元素与响应函数的关联，并包含自动跳转到列表和记录管理界面的功能。下面详细解析该类的各个模块。

1）文件导入的包和类

由于该类需要使用库函数中的模块定义，所以在编写程序时，需 "import" 相应的内容来支持程序中对该类的使用。例如，若使用 Android Studio 中的 "文本" 元件，则必须导入 "TextView" 类，本项目 "MenuActivity.java" 的文件头如下：

```java
package com.as.mininote;

import android.app.Activity;
import android.app.AlertDialog;
import android.content.Context;
import android.content.DialogInterface;
import android.content.Intent;
import android.os.Bundle;
import android.view.KeyEvent;
import android.view.View;
import android.widget.Button;
```

2) MenuActivity.java 文件实现

"MenuActivity.java" 的初始化单元如下：

```java
public class MenuActivity extends Activity {
    private static final int ACTIVITY_CREATE = 0;
    private Context mcontext=this;
    public void onCreate(Bundle savedInstanceState) {
        super.onCreate(savedInstanceState);
        setContentView(R.layout.activity_menu);
        Button allButton = (Button) findViewById(R.id.all);
        Button newButton = (Button) findViewById(R.id.newinsert);
        Button exitButton = (Button) findViewById(R.id.exitp);

        allButton.setOnClickListener(new View.OnClickListener() {
            public void onClick(View view) {
                Intent i = new Intent(mcontext, MainActivity.class);
                startActivityForResult(i, ACTIVITY_CREATE);
            }
        });

        newButton.setOnClickListener(new View.OnClickListener() {
            public void onClick(View view) {
                Intent i = new Intent(mcontext, NoteEditActivity.class);
                startActivityForResult(i, ACTIVITY_CREATE);
            }

        });

        exitButton.setOnClickListener(new View.OnClickListener() {
            public void onClick(View view) {

                AlertDialog.Builder builder=new AlertDialog.Builder(mcontext);
                builder.setTitle("提示");
                builder.setMessage("确定要退出程序?");
                builder.setPositiveButton("确定", new DialogInterface
                    .OnClickListener() {
```

```java
            @Override
            public void onClick(DialogInterface dialog, int which) {
                dialog.dismiss();
                System.exit(0);
                finish();
            }
        });
        builder.setNeutralButton("取消", new DialogInterface
                .OnClickListener() {

            @Override
            public void onClick(DialogInterface dialog, int which) {
                dialog.dismiss();
            }
        });
        builder.show();
    }
});
    }
}
```

3) 手机"回退"按钮响应函数

由于一般 Android 手机都有"回退"按钮，而 Android 程序也应该支持"回退"按钮退出程序这一功能，本项目实现该功能的详细代码如下：

```java
public boolean onKeyDown(int keyCode, KeyEvent event) {

    if(keyCode == KeyEvent.KEYCODE_BACK ){
    AlertDialog.Builder builder=new AlertDialog.Builder(this);
    builder.setTitle("提示");
    builder.setMessage("确定要退出程序?");
    builder.setPositiveButton("确定", new DialogInterface
            .OnClickListener() {
        @Override
        public void onClick(DialogInterface dialog, int which) {
            dialog.dismiss();
            finish();
        }
    });
    builder.setNeutralButton("取消", new DialogInterface
            .OnClickListener() {
        @Override
        public void onClick(DialogInterface dialog, int which) {
            dialog.dismiss();
        }
    });
    builder.show();
```

第16章 Android 数据库记事本

```
        return true;
    }else{
        return super.onKeyDown(keyCode, event);
    }
}
```

2. 实现主程序 MainActivity

1）文件导入的包和类

由于该类需要使用库函数中的模块定义，所以在编写程序时，需"import"相应的内容来支持程序中对该类的使用。MainActivity.java 文件头如下：

```
package com.as.mininote;

import android.app.ListActivity;
import android.content.Intent;
import android.database.Cursor;
import android.os.Bundle;
import android.view.Menu;
import android.view.MenuItem;
import android.view.View;
import android.widget.ListView;
import android.widget.SimpleCursorAdapter;
```

2）实现 MainActivity.java

MainActivity.java 的初始化单元如下：

```
public class MainActivity extends ListActivity {
private static final int ACTIVITY_CREATE = 0;
private static final int ACTIVITY_EDIT = 1;

private static final int INSERT_ID = Menu.FIRST;
private static final int DELETE_ID = Menu.FIRST + 1;

private DiaryDbAdapter mDbHelper;
private Cursor mDiaryCursor;

@Override
public void onCreate(Bundle savedInstanceState) {
    super.onCreate(savedInstanceState);
    setContentView(R.layout.activity_main);
    mDbHelper = new DiaryDbAdapter(this);
    mDbHelper.open();
    renderListView();
}
```

3）界面初始化

该函数的主要功能为调用数据库接口函数，得到数据库中的记录，并将其按行列表到界面上，具体代码如下：

```
private void renderListView() {
```

```java
        mDiaryCursor = mDbHelper.getAllNotes();
        startManagingCursor(mDiaryCursor);
        String[] from = new String[] { DiaryDbAdapter.KEY_TITLE,
                DiaryDbAdapter.KEY_CREATED };
        int[] to = new int[] { R.id.text1, R.id.created };
        SimpleCursorAdapter notes = new SimpleCursorAdapter(this,
                R.layout.activity_note_list, mDiaryCursor, from, to);
        setListAdapter(notes);
    }
```

4) 菜单初始化及响应函数

该工程还提供了菜单操作,用户可以通过菜单项目的选择,进行添加和删除操作。当选择添加时进入编辑界面,具体代码如下:

```java
@Override
public boolean onCreateOptionsMenu(Menu menu) {
    super.onCreateOptionsMenu(menu);
    menu.add(0, INSERT_ID, 0, R.string.menu_insert);
    menu.add(0, DELETE_ID, 0, R.string.menu_delete);
    return true;
}

@Override
public boolean onMenuItemSelected(int featureId, MenuItem item) {
    switch (item.getItemId()) {
        case INSERT_ID:
            createDiary();
            return true;
        case DELETE_ID:
            mDbHelper.deleteDiary(getListView().getSelectedItemId());
            renderListView();
            return true;
    }
    return super.onMenuItemSelected(featureId, item);
}

private void createDiary() {
    Intent i = new Intent(this, NoteEditActivity.class);
    startActivityForResult(i, ACTIVITY_CREATE);
}
```

5) 列表元素单击响应函数

根据本项目功能的设定,在列表界面 MainActivity 中单击其中的列表项,应打开对应记事的编辑界面,在编辑界面中单击"回退"按钮时会回到列表界面,具体代码如下:

```java
@Override
protected void onListItemClick(ListView l, View v, int position, long id) {
    super.onListItemClick(l, v, position, id);
    Cursor c = mDiaryCursor;
```

```
        c.moveToPosition(position);
        Intent i = new Intent(this, NoteEditActivity.class);
        i.putExtra(DiaryDbAdapter.KEY_ROWID, id);
        i.putExtra(DiaryDbAdapter.KEY_TITLE, c.getString(c
                .getColumnIndexOrThrow(DiaryDbAdapter.KEY_TITLE)));
        i.putExtra(DiaryDbAdapter.KEY_BODY, c.getString(c
                .getColumnIndexOrThrow(DiaryDbAdapter.KEY_BODY)));
        startActivityForResult(i, ACTIVITY_EDIT);
    }

    @Override
    protected void onActivityResult(int requestCode, int resultCode,
                            Intent intent) {
        super.onActivityResult(requestCode, resultCode, intent);
        renderListView();
    }
```

3. 实现 NoteEditActivity.java 文件

1) 文件导入的包和类

由于该类需要使用库函数中的模块定义,所以在编写程序时,需 "import" 相应的内容来支持程序中对该类的使用。NoteEditActivity.java 文件头如下:

```
    package com.as.mininote;

    import android.app.Activity;
    import android.app.AlertDialog;
    import android.content.Context;
    import android.content.DialogInterface;
    import android.content.Intent;
    import android.os.Bundle;
    import android.view.View;
    import android.widget.Button;
    import android.widget.EditText;
```

2) 类的定义

NoteEditActivity.java 的初始化单元如下:

```
    public class NoteEditActivity extends Activity {
        private EditText mTitleText;
        private EditText mBodyText;
        private Long mRowId;
        private DiaryDbAdapter mDbHelper;
        private Context mcontext=this;

        @Override
        protected void onCreate(Bundle savedInstanceState) {
            super.onCreate(savedInstanceState);
            mDbHelper = new DiaryDbAdapter(this);
            mDbHelper.open();
```

```java
setContentView(R.layout.activity_note_edit);

mTitleText = (EditText) findViewById(R.id.title);
mBodyText = (EditText) findViewById(R.id.body);

Button confirmButton = (Button) findViewById(R.id.confirm);
Button deletButton = (Button) findViewById(R.id.delet);

mRowId = null;

Bundle extras = getIntent().getExtras();
if (extras != null) {
    String title = extras.getString(DiaryDbAdapter.KEY_TITLE);
    String body = extras.getString(DiaryDbAdapter.KEY_BODY);
    mRowId = extras.getLong(DiaryDbAdapter.KEY_ROWID);

    if (title != null) {
        mTitleText.setText(title);
    }
    if (body != null) {
        mBodyText.setText(body);
    }
}
}
```

3)"保存"按钮响应函数

本项目中的编辑界面最重要的功能是编辑、保存、删除,而编辑功能是文本框自带的功能实现的,但是保存和删除功能则需要操作数据库,这部分是编程的重点。其中"保存"按钮的响应函数完成判断用户输入是否合法,根据输入新建或者更新的内容存入数据库,并且给出用户提示的操作,具体代码如下:

```java
confirmButton.setOnClickListener(new View.OnClickListener() {
    public void onClick(View view) {
        String title = mTitleText.getText().toString().trim();
        String body = mBodyText.getText().toString().trim();
        if(!title.equals("")&&!body.equals("")){
            if (mRowId != null) {
                mDbHelper.updateDiary(mRowId, title, body);
            } else
                mDbHelper.createDiary(title, body);
            AlertDialog.Builder builder=new AlertDialog.Builder(mcontext);
            builder.setTitle("提示");
            builder.setMessage("保存成功");
            builder.setPositiveButton("确定", newDialogInterface
                    .OnClickListener() {
                @Override
                public void onClick(DialogInterface dialog, int which) {
                    dialog.dismiss();
```

```
                finish();
            }
        });
        builder.show();

        Intent mIntent = new Intent();
        setResult(RESULT_OK, mIntent);

    }
    else{
        AlertDialog.Builder builder=new AlertDialog.Builder(mcontext);
        builder.setTitle("提示");
        builder.setMessage("标题与内容不能为空！");
        builder.setPositiveButton("确定", new DialogInterface.
                OnClickListener() {
            @Override
            public void onClick(DialogInterface dialog, int which) {
                dialog.dismiss();
            }
        });
        builder.show();
    }
  }
});
```

4) "删除"按钮响应函数

"删除"按钮的响应函数，需要判断用户删除操作是否合法，合法后定位到记录进行删除数据库记录，并提示用户成功与否。具体代码如下：

```
deletButton.setOnClickListener(new View.OnClickListener() {
    public void onClick(View view) {
        if (mRowId != null) {
            mDbHelper.deleteDiary(mRowId);
            AlertDialog.Builder builder=new AlertDialog.Builder(mcontext);
            builder.setTitle("提示");
            builder.setMessage("删除成功！");
            builder.setPositiveButton("确定", new DialogInterface
                    .OnClickListener() {
                @Override
                public void onClick(DialogInterface dialog, int which) {
                    dialog.dismiss();
                    finish();
                }
            });
            builder.show();
            Intent mIntent = new Intent();
            setResult(RESULT_OK, mIntent);
        }
    }
});
```

4．实现数据库 DiaryDbAdapter.java 类

1）数据库设计

本项目需要使用数据库来存储用户的记事本数据，这里采用的是 SQLite 数据库。SQLite 是一款轻量级数据库，是遵守ACID的关系型数据库管理系统。SQLite 数据库操作相对方便，由两个封装好的类"DatabaseHelper"、"SQLiteDatabase"即可完成几乎所有需要的数据库功能。本例中的数据库设计仅为举例需要，相对简单，只由一个表格完成，数据库设计如表 16.1 所示。

表 16.1 数据库设计

字段名	字段类型	主键	备注
_id	integer	是	编号
title	text	否	记事名称
body	text	否	记事内容
created	text	否	存储时间

2）DiaryDbAdapter 文件导入的包和类

DiaryDbAdapter 是用来包含数据库访问和定义的类，该类同样也包含了其他一些类定义的函数和变量，本项目"DiaryDbAdapter.java"的文件头如下：

```java
package com.as.mininote;

import java.util.Calendar;
import android.content.ContentValues;
import android.content.Context;
import android.database.Cursor;
import android.database.SQLException;
import android.database.sqlite.SQLiteDatabase;
import android.database.sqlite.SQLiteOpenHelper;
```

3）类的定义

构造函数是程序初次运行时自动调用的函数，一般在这里对程序中的变量、界面元素等做生成和初始化，DiaryDbAdapter.java 文件的构造函数定义如下：

```java
public class DiaryDbAdapter {

    public static final String KEY_TITLE = "title";
    public static final String KEY_BODY = "body";
    public static final String KEY_ROWID = "_id";
    public static final String KEY_CREATED = "created";

    private static final String TAG = "DiaryDbAdapter";
    private DatabaseHelper mDbHelper;
    private SQLiteDatabase mDb;
    private final Context mCtx;
    public DiaryDbAdapter(Context ctx) {
        this.mCtx = ctx;
    }
```

第 16 章　Android 数据库记事本

```java
public DiaryDbAdapter open() throws SQLException {
    mDbHelper = new DatabaseHelper(mCtx);
    mDb = mDbHelper.getWritableDatabase();
    return this;
}
```

4）数据库表的创建

记事本需要用到的数据库表必须在程序初始化时创建，而且需要防止重复创建导致数据库返回错误，所以需要在类构造时就进行。这里采取的办法是在成员中声明一个自己重载的数据库连接类，并在该类构造函数中完成数据库表的初始化创建，具体代码如下：

```java
private static final String DATABASE_CREATE = "create table diary (_id integer 
    primary key autoincrement, "+ "title text not null, 
    body text not null, created text not null);";
private static final String DATABASE_NAME = "database";
private static final String DATABASE_TABLE = "diary";
private static final int DATABASE_VERSION = 1;
private static class DatabaseHelper extends SQLiteOpenHelper {

    DatabaseHelper(Context context) {
        super(context, DATABASE_NAME, null, DATABASE_VERSION);
    }

    @Override
    public void onCreate(SQLiteDatabase db) {
        db.execSQL(DATABASE_CREATE);
    }

    @Override
    public void onUpgrade(SQLiteDatabase db, int oldVersion, int newVersion) {
        db.execSQL("DROP TABLE IF EXISTS diary");
        onCreate(db);
    }
}
```

5）数据库的增、删、改、查

本项目中数据库表的增、删、改、查都通过我们重载的数据库连接类来进行，并对父类的对应函数根据程序的需求进行修改，具体代码如下：

```java
public long createDiary(String title, String body) {
    ContentValues initialValues = new ContentValues();
    initialValues.put(KEY_TITLE, title);
    initialValues.put(KEY_BODY, body);
    Calendar calendar = Calendar.getInstance();
    String created = calendar.get(Calendar.YEAR) + "年"
        + calendar.get(Calendar.MONTH) + "月"
        + calendar.get(Calendar.DAY_OF_MONTH) + "日"
        + calendar.get(Calendar.HOUR_OF_DAY) + "时"
```

```java
            + calendar.get(Calendar.MINUTE) + "分";
    initialValues.put(KEY_CREATED, created);
    return mDb.insert(DATABASE_TABLE, null, initialValues);
}

public boolean deleteDiary(long rowId) {

    return mDb.delete(DATABASE_TABLE, KEY_ROWID + "=" + rowId, null) > 0;
}

public Cursor getAllNotes() {

    return mDb.query(DATABASE_TABLE, new String[] { KEY_ROWID, KEY_TITLE,
            KEY_BODY, KEY_CREATED }, null, null, null, null, null);
}

public Cursor getDiary(long rowId) throws SQLException {

    Cursor mCursor =
            mDb.query(true, DATABASE_TABLE, new String[]
                    { KEY_ROWID, KEY_TITLE, KEY_BODY, KEY_CREATED },
                    KEY_ROWID + "=" + rowId, null, null,
                null, null, null);
    if (mCursor != null) {
        mCursor.moveToFirst();
    }
    return mCursor;
}

public boolean updateDiary(long rowId, String title, String body) {
    ContentValues args = new ContentValues();
    args.put(KEY_TITLE, title);
    args.put(KEY_BODY, body);
    Calendar calendar = Calendar.getInstance();
    String created = calendar.get(Calendar.YEAR) + "年"
            + calendar.get(Calendar.MONTH) + "月"
            + calendar.get(Calendar.DAY_OF_MONTH) + "日"
            + calendar.get(Calendar.HOUR_OF_DAY) + "时"
            + calendar.get(Calendar.MINUTE) + "分";
    args.put(KEY_CREATED, created);
    return mDb.update(DATABASE_TABLE, args, KEY_ROWID + "=" + rowId, null) > 0;
}
```

16.4 系统运行与测试

界面设计和代码设计完成后，单击 Android Studio 开发环境的工具栏中的 ▶ 图标，或者选择菜单栏中的"Run"→"Run"命令，运行该项目，可以调试查看 Android 数据库记事本

第 16 章　Android 数据库记事本

程序。当我们单击运行后，会弹出设备选择窗口，如图 16.13 所示，这里可以选择连接在计算机上的手机设备、运行中的 Android 虚拟机或者新启动一个 Android 虚拟设备。这里选择启动早先使用 AVD 创建好的"Nexus S"设备。

图 16.13　设备选择窗口

在运行好的窗口中增加记录，然后返回列表查看，最后删除记录，所有步骤成功，项目开发完成！如图 16.14 所示。

图 16.14　项目运行结果——记录增、查、删的过程

参 考 文 献

[1] 李刚. 疯狂 Android 讲义（第 3 版）. 北京：电子工业出版社，2015.
[2] 郭霖. 第一行代码 Android. 北京：人民邮电出版社，2014.
[3] [美] Reto Meier 著；佘建伟，赵凯译. Android 4 高级编程（第 3 版）. 北京：清华大学出版社，2013.
[4] 明日科技（中国）有限公司. Android 从入门到精通. 北京：清华大学出版社，2012.
[5] 向守超，姚骏屏. Android 程序设计实用教程. 北京：电子工业出版社，2013.
[6] 王友钊，黄静，戴燕云. Android 系统开发与实践. 北京：清华大学出版社，2013.
[7] 刘超. 深入解析 Android 5.0 系统. 北京：人民邮电出版社，2015.
[8] 熊庆宇. 软件工程实训项目案例——Android 移动应用开发篇. 重庆：重庆大学出版社，2014.
[9] 林学森. 深入理解 Android 内核设计思想. 北京：人民邮电出版社，2014.
[10] [美] Carlos Sessa 著；杨云君译. 打造高质量 Android 应用：Android 开发必知的 50 个诀窍. 北京：机械工业出版社，2014.
[11] [瑞典] Erik Hellman 著；丁志虎，武海峰译. Android 编程实战. 北京：人民邮电出版社，2014.
[12] 武永亮. Android 开发范例实战宝典. 北京：清华大学出版社，2014.
[13] 陈文，郭依正. 深入理解 Android 网络编程：技术详解与最佳实践. 北京：机械工业出版社，2013.
[14] 郭金尚. Android 经典项目案例开发实战宝典. 北京：清华大学出版社，2013.
[15] [印] Satya Komatineni，[美] Dave MacLean 著；曾少宁，杨越译. 精通 Android. 北京：民邮电出版社，2011.
[16] 佘志龙. Google Android SDK 开发范例大全. 北京：人民邮电出版社，2011.
[17] 邓凡平. 深入理解 Android：Wi-Fi、NFC 和 GPS 卷. 北京：机械工业出版社，2014.
[18] 李钦. 基于工作项目的 Android 高级开发实战. 北京：电子工业出版社，2015.
[19] 姚尚朗，靳岩. Android 开发入门与实战（第 2 版）. 北京：人民邮电出版社，2013.
[20] 软件开发技术联盟. 软件开发实战：Android 开发实战. 北京：清华大学出版社，2013.

反侵权盗版声明

电子工业出版社依法对本作品享有专有出版权。任何未经权利人书面许可，复制、销售或通过信息网络传播本作品的行为；歪曲、篡改、剽窃本作品的行为，均违反《中华人民共和国著作权法》，其行为人应承担相应的民事责任和行政责任，构成犯罪的，将被依法追究刑事责任。

为了维护市场秩序，保护权利人的合法权益，我社将依法查处和打击侵权盗版的单位和个人。欢迎社会各界人士积极举报侵权盗版行为，本社将奖励举报有功人员，并保证举报人的信息不被泄露。

举报电话：（010）88254396；（010）88258888
传　　真：（010）88254397
E-mail：dbqq@phei.com.cn
通信地址：北京市海淀区万寿路 173 信箱
　　　　　电子工业出版社总编办公室
邮　　编：100036